MOUNTAINS OF MADNESS
A Scientist's Odyssey in Antarctica

John Long

Foreword by Tim Bowden

JOSEPH HENRY PRESS
Washington, D.C.

Joseph Henry Press • 2101 Constitution Avenue, N.W. • Washington, D.C. 20418

The Joseph Henry Press, an imprint of the National Academy Press, was created with the goal of making books on science, technology, and health more widely available to professionals and the public. Joseph Henry was one of the founders of the National Academy of Sciences and a leader of early American science.

Any opinions, findings, conclusions, or recommendations expressed in this volume are those of the author and do not necessarily reflect the views of the National Academy of Sciences or its affiliated institutions.

Library of Congress Cataloging-in-Publication Data

Long, John A., 1957-
Mountains of madness : a scientist's odyssey in Antarctica / John Long ; foreword by Tim Bowden.
p. cm.
Includes bibliographical references (p.).
ISBN 0-309-07077-5 (alk. paper)
1. Long, John A., 1957—Journeys—Antarctica—Transantarctic Mountains. 2. Paleontology—Antarctica—Transantarctic Mountains. 3. Scientific expeditions—Antarctica—Transantarctic Mountains. 4. Transantarctic Mountains (Antarctica)—Discovery and exploration. I. Title.

QE22.L755 A3 2001
919.8'904—dc 21

00-064494

Cover photograph by John G. McPherson.

Endpapers: Two of the author's maps from the 1991-92 expedition

Published in Australia by Allen & Unwin Pty Ltd

Printed in the United States of America.

For Margaret Bradshaw,

one of the great heroines

of modern Antarctic exploration

Contents

Foreword vii

Preface xi

1 A Strange and Hostile Land 1

2 Of Heroes, Rocks, and Fossils 12

3 Survival Training: Tekapo, New Zealand 22

4 Arrival in Antarctica 29

5 Antics on Ice 39

6 A Very Good Christmas 45

7 Dry Valleys, Sand Dunes, and Rivers 57

8 Back to the Great White South 67

9 Cape Evans and Cape Royds 75

10 A Flight of Discovery 84

11 On Mr. Darwin's Glacier 88

12 The First Worst Day of My Life 91

13 Dancing on the Gorgon's Head 102

14 Up the McCleary Glacier 110

15 A Room at the Fish Hotel 116

16 The Skua 127

17 The Ascent of Mt. Gudmundson 131
18 Over the Mulock Glacier 137
19 Mt. Ritchie and Deception Glacier 142
20 Onwards to Escalade Peak 149
21 "I'm Dreaming of a White Christmas" 155
22 On the Snout of the Alligator 168
23 Long Day's Journey into Night 172
24 Skating Away on the Thin Ice of a New Year 181
25 Working on the Portal 188
26 At the Crucible of Shark Evolution 193
27 Pick-Up Day Problems 203
28 Base Blues and Arrival Home in Australia 207
29 So Much for the Afterglow 213
30 Reflections from the Ice 219
Epilogue 223
Appendix 1 What I Liked and Disliked About Antarctica 227
Appendix 2 Some Great Recipe Ideas from Antarctica 229
References and Source Material 231
Acknowledgments 235
Index 237

Foreword

Tim Bowden

Strange things happen to those who go to Antarctica. There is a strong sense of unreality about confronting nature on such a vast scale. No one who goes there, even for quite short periods, is unaffected by the experience. Phillip Law, who directed Australia's exploration and scientific program from 1949-66, puts it this way:

> It is fair to say that no man ever goes to Antarctica without its having an immense impression on him. There's the beauty and grandeur of it all and the feeling that one is so insignificant in this scale of nature. The fact that you have enough time down there to sit and ponder and philosophize and sort yourself out enables you to look back on civilization from a stand-off point of view . . . and I think most men in Antarctica for any length of time do go through some sort of personal reassessment.

For the scientists and support personnel who work there, Antarctica exerts a compelling fascination, leading to chronic recidivism for many. Yet no humans live permanently in Antarctica. We go there like astronauts into deep space, taking everything needed to sustain life: food, fuel, and accommodation. Twenty-six nations now maintain permanent stations in Antarctica. Scientific research by paleontologists, geologists, meteorologists, glaciologists, upper atmosphere physicists, biologists, and microbiologists—among other professionals—is the driving force for all Antarctic activity, most of which is compressed into the short summer season.

Those hardy souls who venture out into the field from the security of the stations do so at considerable risk. Although they have the advantage of satellite communications and motorized over-snow transport, no crevasse detector has yet been invented, sudden blizzards can overwhelm individuals unexpectedly and frostbite and wind chill are still as vicious as they were to Scott and Mawson. At even greater risk are those whose projects take them into "deep field," more than 200 kilometers from base and outside helicopter rescue range, where the thin fabric of polar tent is all that separates those sheltering inside from the howling super-chilled winds, from life or death. At those times "deep field" might as well be "deep space," with rescue impossible if the worst should happen. This is adventuring and scientific exploration on the cutting edge.

I have not read a finer account of a modern Antarctic field trip than *Mountains of Madness: A Scientist's Odyssey in Antarctica*, by Australian paleontologist John Long. This book chronicles two summer expeditions to Antarctica but principally the deep field trip to the remote Cook Mountains (part of the Transantarctic Mountain chain that separates Greater and Lesser Antarctica) in 1991-92. Long and his companions were searching for fish fossils from the Devonian period some 400 million years ago, first discovered high in the mountains by veteran Antarctic geologist Margaret Bradshaw two years previously. Two men and two women (including Margaret) were flown in by C-130 ski-equipped Hercules aircraft and landed on the ice among the mountains, together with their skidoos, sledges, tents, climbing gear, rations, and geological equipment in virgin territory. The planes were not scheduled to return for eight weeks. Many of the peaks around them had not been climbed or explored.

Long, a scientist-romantic, is a keen student of Antarctic writing and his quotations from men like Byrd, Cherry-Garrard, Mawson, Scott, and Shackleton add luster to this outstanding account. By coincidence, the party landed near the mythical location chosen by novelist H.P. Lovecraft for his classic Gothic story, *At the Mountains of Madness*, where a geological expedition stumbles across a strange and hidden civilization. While blizzard-bound in their tents, the four expeditioners passed the time by reading Lovecraft out loud. Their isolation engendered their own moments of madness, courageously described by Long.

Scientifically, the expedition was a great success, with many new species recovered and named, together with important evidence of how certain life forms in the great conglomeration of Gondwana were later carried by continental drift to distant parts of the globe. Long's great skill is to take the reader through the adventure of a deep field experience while sharing the excitement of the fossil finds and, at the same time, putting it all in historical context.

Phillip Law's observations about no one going to Antarctica without some measure of self-assessment is an understatement when applied to John Long. The combination of near-death experiences with crevasses and avalanches, professional and physical achievement, camaraderie in the field, and the intensity of experiencing Antarctica at its wildest and most primal has, by his own admission, completely reshaped his life. You will have to read this singular account to discover how and why.

There is a clue in one of H.P. Lovecraft's quotes selected by John Long from *At the Mountains of Madness*:

> Half-paralyzed with terror though we were, there was nevertheless fanned within us a blazing flame of awe and curiosity which triumphed in the end.

Tim Bowden is the author of two books on Antarctica, Antarctica and Back in Sixty Days *and* The Silence Calling: Australians in Antarctica 1947-97.

Preface

No one goes to Antarctica without coming back a different person. A little part of Antarctica grows inside you and moulds your character, for better or worse, from the first day you step foot on that frozen land. I never suspected that going to Antarctica would change my life in dramatic ways, but it certainly did. Although many other scientists and adventurers have spent much more time on the continent, in far worse conditions than I ever faced, I believe my story is one of special interest because of the reasons our expedition went to such a remote and hostile place. We went there with a mission to search for the fossil remains and traces of Antarctica's earliest flourishing communities. Antarctica has had a profound effect on my life from then on.

This book started out as an account of what occurred on my two trips to Antarctica. On my second trip during the 1991-92 season, I remember being on a permanent high of either discovery or adventure. The thrill of discovery came whenever we found expansive pavements of rock bristling with 380-million-year-old bones; the excitement of adventure came when we were sledging carefully through hidden crevasse fields, scaling ice cliffs wearing crampons and climbing harnesses, or being almost buried by an avalanche. That's how this book started. I then incorporated some of my own ideas and personal philosophy, which I now see as having been formed mostly from my Antarctic experiences.

In the banks of one's memory the two entities of discovery and adventure combine as one stream of consciousness, as all adventure is merely an act of discovery about one's self; particularly one's inner nature under stress. All scientific discovery is simply revelation about the nature of all things outside of the realm of self, where nature can be measured, quantified, and in some small way, qualified. By coalescing the two, one can experience true discovery, a heightened awareness about the world as we perceive it with our senses and feel it in our hearts. Apsley Cherry-Garrard, a veteran of Scott's 1910-13 Terra Nova expedition, expressed similar sentiment by saying that "exploration is the physical expression of the intellectual passion."

During the three months I spent in Antarctica on the 1991-92 expedition, every evening my colleagues and I took turns at reading aloud a few pages of H.P. Lovecraft's classic gothic story, *At the Mountains of Madness*, first published in 1931. In that book an expedition to the remote wilds of Antarctica finds evidence of an incredibly ancient civilization still inhabited by strange beings. At the time of writing it, Lovecraft based his hidden civilization in a totally remote, unexplored location inland from the Transantarctic Mountains at 76° south. This position (which incidentally Lovecraft cites with an exact latitude and longitude) coincided with being directly inland towards the polar plateau from the mountains where we were based near the end of our long sledging journey. Most of the Cook Mountains had never before been investigated on the ground, so we were the first humans to scale several of these lofty peaks, to explore for geological treasures and to discover many new fossil sites there.

Yet, always, each evening, we would anxiously wait to hear more of Lovecraft's gradually unnerving tale. Recently I pulled out my old maps from the field trip and saw that I had annotated the northern-most mountain range in the Cook Mountains as the "Mountains of Madness." The name has no official status. It was our little joke for the expedition, yet somehow it has stuck in the back of my mind.

My career as a paleontologist has taken me to many interesting places to collect or study fossils over the last decade—throughout Australia, Europe, North America, South-East Asia, South Africa, Iran, and Antarctica. Never in my wildest dreams as a child did I imagine that the study of long-dead fishes (my particular specialty) would lead me to so many fascinating places, usually way off the beaten track where tourists never venture. Yet of all the places I have visited, Antarctica was without any doubt the most scenic, dangerous, and scientifically interesting of the lot. Moreover, it left an indelible impression on my soul. This book is my personal story of Antarctica, told from a slightly different perspective to those of regular visitors to the south, as it is an account of two scientific expeditions as well as the ensuing story of how my life changed from those experiences.

One of the main aims of our expeditions was to discover new information on Antarctica's ancient creatures, as only by filling in more of the blanks in Antarctica's bleak fossil record can we eventually hope to solve the larger problem of how the global biodiversity of this planet unfolded. Antarctica's ancient prehistory may hold the key to how and why many of the animals and plant communities around the world today came to be where they are. Furthermore, I believe that if we read carefully between the lines, the story of planet Earth's rich prehistory may give some rare glimpses into where we as a species, in a complex ecosystem, might be heading in the future. You can read more about the prehistoric life of Antarctica in another book of mine, *Life Frozen in Time*, which will be published by Johns Hopkins University Press.

Some excerpts from the writings of the early heroic explorers have been included throughout this story to add color and historical content to the places we visited and the similar situations in which we found ourselves. There was no intention here of making direct comparison between our "comfortable" modern expeditions and the rigors and suffering of the early expeditioners who "man-hauled" most of their way to their destinations.

The stories of the expeditions of Scott, Shackleton, and Mawson are referred to throughout this narrative, for reasons that are clear: our journey started off from the same base area on Ross Island as did Scott's and Shackleton's expeditions, and I was fortunate to visit their original huts. Mawson was way over the other side of Antarctica from where I was, but as his story is one of the best, and as he was a geologist like myself, I have liberally drawn from his tale to illustrate certain points. Similarly, I have used quotes from the writings of some of the later explorers, like Admiral Richard Byrd, an American who was the first man to winter alone on the barrier ice of Antarctica in 1934, and of others whose writings show the uncanny ability to express similar feelings to those I experienced with a clarity that I could not hope to emulate.

The heroic deeds and accomplishments of these brave men are to be highly lauded, and I hope that you gain a sense of admiration for them. I have come to appreciate their deeds, not just from reading their accounts, but also because I have been privileged to visit the same places as some of them.

And, by some miracle of good fortune, lived to tell the tale.

John Long
September 2000

MOUNTAINS OF MADNESS

1
A Strange and Hostile Land

I am forced into speech because men of science have refused to follow my advice without knowing why. It is altogether against my will that I tell my reasons for opposing this contemplated invasion of the Antarctic—with its vast fossil hunt and its wholesale boring and melting of the ancient ice caps.

—H.P. Lovecraft

Thus begins the epic horror story *At the Mountains of Madness*, a twisted tale of an Antarctic fossil-hunting expedition that went horribly wrong, culminating in the death or madness of most of the expeditioners. At the time this was being written, in 1930, Sir Douglas Mawson was on his last Antarctic expedition, the British-Australian-New Zealand Antarctic Research Expedition (BANZARE 1929-31), cruising and mapping much of the unseen coastline of this vast, largely unknown continent. In those days it would not have seemed so mysterious for a continent the size of Antarctica to hold many scientific secrets, perhaps even the vestigial traces of lost civilizations or the fossil remains of higher life forms not found anywhere else on the planet, as Lovecraft's doomed explorers eventually discovered. The reality was simple in 1931: an incredibly small portion of the Antarctic landmass had been explored at all, and virtually nothing was known of its geology or its paleontology.

Most people, quite rightly, think of Antarctica as a land of hostility, an almost alien and unfriendly landscape. Mawson dubbed it the "home of the blizzard" in his epic book of the same name. Vaughan Williams' musical score for the 1948 classic film *Scott of the Antarctic* (which became his *Sinfonia Antarctica* in 1953) used haunting, lilting tones to paint a musical portrait of a cruel, inhuman Antarctica; his refrains are somewhat similar to how one imagines the mythological sirens' songs that were said to lure sailors to their deaths.

Yet Antarctica is actually a land of such unpredictability and natural fury that humans are almost as out of place there as on the moon. On the moon a person requires total protection from the vacuum and cold of space, plus warmth and air to live, and must take all the necessary food and liquid for survival. Similarly, in the coldest parts of Antarctica, say in a midwinter blizzard on the polar plateau, a thousand kilometers* away from the subzero seas, blasting cold winds can freeze human flesh solid in less than a few seconds. The fragile human body must be completely protected by many layers of special clothing, covering all exposed parts of the body. All food must be taken along, and much energy is required to melt ice and snow for drinking water. Like the astronauts on the moon, all human waste is generally taken away as part of a program for protecting a pristine natural environment. Without either a spacesuit on the moon, or heavy clothing and shelter in a midwinter Antarctic blizzard, the human body would rapidly succumb to the extreme conditions.

The discovery of Antarctica by humans dates back just over the last 200 years or so, although Antarctica's geological history goes back at least 3900 million years, based on the radiometric dates of the formation of the mineral zircon in granulitic rocks from Enderby Land. At that time, in the early part of the Archaean Eon, the Earth's crust was still very hot and the first rocks had just cooled enough to form a thin crust. The oldest rocks known from Antarctica are dated at around 3100 million years. At around 2450 million years ago, these older rocks, now forming the original crust of Antarctica, were strongly folded and broken apart by violent intrusions of molten magma. We have no accurate picture of what the Earth looked like at this time, so have no concept of crustal plates with definable boundaries, such as we would call "continents."

---

*Meters and kilometers are the units of measurement used throughout this book. A meter is about three feet and three inches and a kilometer equals about six-tenths of a mile.

Antarctica became a recognizable continental region at least 500 million years ago. At around this time the western and eastern parts of Antarctica are thought to have collided, forming the main landmass we know as Gondwana, the giant southern supercontinent. The name "Gondwanaland" (meaning "land of the Gonds", a native tribe from India) was coined by E. Suess, an Austrian geologist who intuitively recognized geological similarities between the southern continents and peninsular India. Suess suggested that these landmasses were joined by land bridges that had since sunk beneath the seas, Atlantis-style. Today we know this is not true as our knowledge of the Earth's processes tells us that the continents, which are merely slabs of the Earth's crust up to 150 kilometers thick, have been slowly moving, possibly pushed by convection currents within the Earth's mantle (the solid rock layer below the crust). Australia, for example, is currently moving at about six centimeters per year northwards, steaming along ready to collide and become part of Asia in about 50 million years' time. No immigration policies needed then.

Gradually, over the last 500 million years, since the very dawn of Antarctica's formation, large chunks of continental crust have rifted away from its margins to become the continents we recognize today. North America, although there is still controversy about its being a part of Gondwana, was probably the first to rift away (about 490 million years ago), followed soon after by parts of Europe, the Middle East and Arabia, then South America, Africa, and India.

Then, about 130 million years ago, there was only Antarctica, Australia, and New Zealand left, forming the now much reduced Gondwana. Australia began its rifting away from Antarctica about 110 million years ago but the two landmasses did not break their crustal umbilicus until about 65 million years ago, at the end of the Cretaceous Period. The formation of these two continents as separate entities approximates very closely the time when the dinosaurs were in their death throes. However, because of long trailing pieces of Australia's crust, like the Lord Howe Rise, the complete rift away from Antarctica was not properly established until the late Eocene or early Oligocene epochs, perhaps as recent as 30 million years ago. Antarctica and Australia are therefore the youngest of all continents. At around 30 to 40 million years ago the circumpolar current developed, initiating the great cooling of the southern continent. The ice caps that now cover more than 95 percent of Antarctica's landmass started forming about 20 million years ago in the midst of the continent. The last forests on the mainland there slowly vanished until none were left around six million years ago.

In 1931 the concept of "plate tectonics" was indeed a controversial and much misunderstood theory that few scientists took seriously. The German geographer Alfred Wegener had published his book on the subject in the 1920s but he never gained acceptance in the scientific circles of his day. In Lovecraft's novel *At the Mountains of Madness*, the summary presented is indeed a futuristic, yet uncannily correct, view of the situation as we understand it today. On finding the maps and charts of the "Ancient Ones," the civilization that is discovered existing in the remote part of the continent, the expedition's scientist (also the narrator) notes how Antarctica was depicted as being the center of a once gigantic supercontinent:

> As I have said, the hypothesis of Taylor, Wegener, and Joly that all continents are fragments of an original Antarctic land mass which cracked from centrifugal force and drifted apart over a technically viscous lower surface—an hypothesis suggested by such things as the complementary outlines of Africa and South America, and the way the great mountain chains are rolled and shoved up—receives striking support from this uncanny source.

Antarctica is also the most recent of the continents to be discovered, explored, and inhabited by us humans. Over 2000 years ago the Greek philosopher Aristotle hinted at the early existence of an unseen southern continent because he saw a need to balance out the mass of the large northern hemisphere continents. As the northern landmasses were under the star *Arktos*, so Aristotle postulated that a great southern land must exist, which he dubbed *Antarktos*. Ptolemy (150 AD), the Egyptian geographer, went further by agreeing that this southern land, which he referred to as *terra australis incognita*, must exist, and furthermore that it would be fertile and populated. He claimed that it was cut off from the rest of the world by a region of fire and some others went on to say that fearful monsters inhabited it. Such ideas naturally discouraged further exploration to the Antarctic region for many centuries to come.

The mythological background to Antarctica as a potentially prosperous and populated land was first properly dispelled by Captain James Cook, who was the first to sail into the Antarctic Circle, below 66° south latitude, on 17 January 1773. His exact words on this momentous occasion are rather droll and uninspiring: "I continued to stand to the south, and on the seventeenth, between eleven and twelve o'clock, we crossed the Antarctic Circle in the longitude of 66 degrees 36 minutes 30 seconds south."

Cook eventually crossed the Antarctic Circle three times in circumnavigating the polar seas. Although he could not confirm that there was no continent within these icy seas, he expressed the notion that land must have been nearby.

No human being had even laid eyes on Antarctica until 1819, when the first person to do so was probably a Russian admiral, Fabian von Bellinghausen, on the Russian Antarctic expedition. Others claim it could have been the American Nathaniel Palmer, who was there at the same time and actually met up with Bellinghausen. Yet it was the lure of biologically rich seas, the temptation of unlimited numbers of whales, seals, and penguins that tempted bold mariners to risk death by venturing into its icy realm. Despite the sighting of Antarctica in 1819, it was to be another eight decades or so later before anyone was able to successfully set foot on the continent.

Norwegian businessman Henryk Bull, a resident of Melbourne, had the idea that Australia should be the first country to set up a whaling industry in the Antarctic seas. He persuaded a retired old whaler, Svend Foyn, to finance an expedition down south in search of whales. They set sail in September 1894 from Melbourne with a small crew in the refurbished whaling ship *Antarctic*. Amongst their party was surveyor and amateur naturalist Carsten Borchgrevink, who later went on to become one of the first men to spend a winter in Antarctica in 1899. On the expedition of the *Antarctic* they became the first people to set foot on the Antarctic mainland. They landed at Cape Adare on 24 January 1895. Bull writes in his book, *Cruise of the Antarctic* (1896):

> January 24th, 1895. Cape Adare was made at midnight. The weather is now favorable for a landing, and at 1am, a party including captain, second mate, Mr. Borchgrevink, and the writer, set off, landing on a pebbly beach of easy access, after an hour's rowing through loose ice, negotiated without difficulty.

From that point on, the story of Antarctica is much better known through its era of heroic exploration. The names of Scott, Shackleton, Mawson, and Amundsen, the big four, as they are commonly referred to, are as well known as their legendary exploits.

Robert Falcon Scott, naval officer, led the first official British attempt to get to the South Pole on the 1901-04 Discovery expedition. Ernest Shackleton, an Irishman, pulled a sledge alongside Scott and naturalist Edward Wilson on this trip, which resulted in them arriving at 82° south, although Shackleton suffered from scurvy and later had to be invalided out. With a burning desire to prove himself, Shackleton went on to lead his own assault on the pole on the 1907-09 British Antarctic expedition, and his team reached furthest south at 88°23′ south on 9 January 1909. This same expedition claimed the honor of being the first to reach the South Magnetic Pole. It included a young English geologist who had been living in South Australia, Douglas Mawson, and a well-known geologist who worked at the University of Sydney, Professor Edgeworth David. These two intrepid explorer-scientists were also part of a team that climbed the active volcano Mt. Erebus for the first time.

Scott's most famous and ill-fated trip, the 1910-13 Terra Nova expedition, is well documented by many good books, films, and even a British television mini-series. The heroic efforts of Scott's team's long march to the pole and their tragic deaths on the way back made the world headlines of the day, practically overshadowing the success of Norwegian Raold Amundsen's team in being the first men to reach the South Pole. Amundsen and his men had made clever use of many dogs to race to the pole and get there with almost clinical precision on 14 December 1911. By using some of the dogs as food for the other dogs, the work of the sledge hauling was greatly minimized for his men. Scott, on the other hand, had brought ponies and cars to lug supplies out to his depots. The cars, looking more like snow tractors, eventually packed up from overheating and other unforeseen mechanical problems and the ponies, although holding out well despite the severe conditions, could not get up past the Beardmore Glacier. While it cannot be fairly said that Scott didn't plan his march well, or try very hard, they were dogged with many problems along the way and had gotten off to a dangerously late start. Eventually, it was to be the unpredictability of the cruel Antarctic weather, particularly the fierce storms, that would prevent them from reaching One Ton Depot, only 19 kilometers away. The three remaining men, Scott, Wilson, and Bowers, perished in their tent and were not found until the following season.

Douglas Mawson led the Australasian Antarctic expedition from 1911 to 1914, with the principal aim of mapping and carrying out scientific work in Antarctica. Through his efforts on this expedition and the 1929-31 British-Australian-New Zealand Antarctic Research Expedition (BANZARE), he is today credited with having made Australia's claim to almost 42 percent of the Antarctic continent, some six million square kilometers of land! The story of Mawson's lone trek back to the Commonwealth Bay hut after the accidental deaths of his two companions, Ninnis and Mertz, has to rate as one of the most incredible tales of human endeavor, pure courage and raw strength. I make no hesitation in stating right here that Mawson is my favorite of all the heroic explorers, mainly because he was a scientist who made these gallant efforts for the advancement of science; yet his bravery in getting his work done surely must rate up there with that of any of the other early polar explorers.

Today Antarctica still holds much mystery and inspires much awe due to its extreme and unpredictable climate and the fact that it is a vast continent, most of which has never been explored on the ground by humans. While it is true that the greater part of it comprises the boundless polar plateau of ice and snow, even the Transantarctic Mountains belt has only been explored in select regions, mostly those areas easily reached from the bases that fringe the continent. To understand the nature of how difficult it is to get to remote parts of Antarctica, one need only look at the map shown in this book and the scale of the continent.

Whereas a limited number of bases with helicopters can give access to destinations within a few hundred kilometers of each base, anywhere else on the continent can only be reached by a major plane flight. Today the US Antarctic program (USAP) uses a fleet of C-130 Hercules aircraft, specially modified for work in Antarctic conditions for this purpose. Through this scheme of logistics both American and New Zealand expeditions have gained rare access to the most remote inland parts of the continent to explore and sample its long-guarded scientific secrets.

Such expeditions are dubbed "deep field" because they are out of helicopter reach (more than a 200-kilometer radius from the base) and must rely on being put in and pulled out by Hercules aircraft. Thus, the expeditioners are reliant solely on their radio contacts with the base to plot positions and for an aircraft to be available to pick them up at the end of the expedition. Sometimes this is not so simple as by the end of the season planes are often out of action due to mechanical problems, or weather can turn sour at very short notice, and so field parties that have completed their work may have to wait patiently for several weeks just to be picked up.

I had always wanted to go to Antarctica and search for fossils ever since hearing the tales of early fossil-hunting expeditions recounted to me by my good friends Dr. Alex Ritchie (now at the Australian Museum) and Dr. Gavin Young (at the Australian National University). Both are specialists who study the fishes of the Devonian period (355-408 million years ago), an age when the evolution of fishes was probably the most exciting thing happening around the world—when fishes first evolved into land animals, conquering amazing hardships to leave the water and invade the land.

Alex and Gavin had been on an expedition to the Skelton Névé region in 1970-71 with a group from the Victoria University of Wellington (hence named VUWAE expedition 15) and had been the first scientists to systematically collect fossils from the rocky outcrops exposed in the region. There are now some 30 different species of fossil fishes described from the Devonian age rocks exposed in southern Victoria Land. This research has enabled strong correlations to be made between the fossil fish fauna of Antarctica and those of similar age in Australia, South Africa, and other Gondwana countries.

Unfortunately for me, the Australian bases were not close to any of the fossil sites I was interested in visiting. During the course of my doctoral work I, too, had become hooked on studying the fishes of the Devonian period, so I desperately wanted to continue the search for new fossil sites in Antarctica. The previous sites visited by Alex Ritchie and Gavin Young, some 20 years earlier, had produced many splendid fish fossils, yet only in the late 1980s had New Zealand field parties discovered that these same fossil-bearing rocks (called the "Aztec Siltstone" after a pyramidal mountain that somewhat resembled an Aztec temple) actually extended at least 200 kilometers further south than previously thought. These new exposures in the Cook Mountains were unexplored, so the potential for major new discoveries was very high indeed.

Margaret Bradshaw, a veteran of Antarctic geology, first made this discovery in the 1988-89 field season. I was stuck back at Scott Base at the time anxiously hoping to join the last leg of her trip to collect fossils while she and her colleagues were out making their discoveries in the southern part of the Cook Mountains. Unfortunately, I didn't get to these fossil sites on that first trip so a much larger expedition had to be planned, which would take in all the possible mountain ranges where the newly discovered exposures of the Aztec Siltstone should occur. This meant going into a part of the Transantarctic Mountains into which no humans had previously ventured. We would have to sledge up glaciers for the first time, climb mountains for the first time and, hopefully, discover many new fossil sites. So what got the whole thing kick-started in the first place?

In 1985 I attended an Australasian paleontological meeting in Christchurch, the Hornibrook Symposium—named in honor of a famous New Zealand paleontologist. I visited the Canterbury Museum and examined the superb fish fossils that Margaret Bradshaw, then Curator of Geology at the Canterbury Museum, had collected on her various trips to Antarctica. Margaret came across as a headstrong, physically fit woman, clearly one of the new breed of fearless female Antarctic explorer-scientists. She always wore a somewhat cheeky grin, had a frizzled-out hairstyle and maintained that kind of cheery enthusiasm for her work that would make most nine-to-fivers sick. Raised in

Nottingham, England, she moved to Christchurch, where she carved a niche for herself in New Zealand as the expert in Devonian invertebrate paleontology and eventually moved on to looking at similar fossil assemblages occurring in Antarctica. By 1985 she had been to Antarctica several times, mostly on deep field expeditions. Her husband, John Bradshaw, a lecturer at Canterbury University, is also a veteran of many trips studying geology down in Antarctica.

Women were excluded from working or participating in any Antarctic expedition right up until almost the mid-twentieth century. The old notion of Antarctica being "no place for a woman" went without saying in those days. It wasn't until 20 February 1935 that Caroline Mikkelsen, a Norwegian, became the first woman to step foot on the frozen continent. In 1946 two American pilots, Harry Darlington and Finne Ronne, took their wives to Stonington Island on a spur of the moment decision, but not without some dissent from the other men who signed a petition trying to stop it happening. In the end both wives accompanied their husbands on the southern cruise of *The Port of Beaumont*, thereby becoming the first women to winter over in Antarctica. Jennie Darlington wrote a book on her experiences entitled *My Antarctic Honeymoon*, published in 1956.

However, it wasn't until 1957, in the International Geophysical Year, that Russia became the first country to take a woman down to Antarctica to work there. This was the first time a woman went to Antarctica who actually wasn't married to an expeditioner! Australia brought its first woman to Antarctica shortly afterwards, in 1959, but the Americans didn't follow suit until ten years later. Today, I'm pleased to say, there is a large proportion of women active in scientific research programs or working around the bases in Antarctica, although the gender balance is still nowhere near equitable.

Margaret Bradshaw's particular field of expertise was the study of fossil shells of the Devonian period (particularly bivalves or clams), but she had also turned her attention to investigating the trace fossils (the fossilized remains of animals burrows, footprints or feeding trails) and sedimentary geology of the older Devonian sequences in Antarctica to reconstruct its ancient environments.

So it was on that day in December 1985, while we were working in the Canterbury Museum, that she told me that she was going back down to Antarctica, and asked me casually: "Why don't you come along as the fish fossil expert on our field party?" I jumped at this opportunity and we put the proposal to the various funding bodies, both through the New Zealand Antarctic Research Program (NZARP) in New Zealand and the National Geographic Society of America. Both proposals were approved soon after, so Margaret had the field party and logistics organized, and I had the funding from the National Geographic Society to participate from Australia.

The burning question that everyone asks me about Antarctic fossils is: where do you find them? Do you dig holes in the snow for them? Some have even asked me if the animals were frozen in ice, like the famous cases of the woolly mammoths of Siberia, trapped snap-frozen by a sudden snowstorm. The true answer is that fossils occur in layers of sedimentary rocks, and these are exposed in the mountainous regions of Antarctica simply because these are the only areas not covered by a thick layer of ice.

It is interesting to follow the history of the early fossil discoveries in Antarctica, as all of the main expeditions were set up primarily for scientific purposes. Roald Amundsen, who just wanted to be the first person to reach the South Pole, was perhaps the exception. Nevertheless, his men still made various observations on the geology, climate, and magnetism along the journey and when they finally reached the pole made some useful scientific measurements. Edward Wilson, scientist on both of Scott's expeditions, said before the second attempt on reaching the South Pole: 'We want the Scientific work to make the bagging of the Pole merely an item in the results.'

However, when seen from the perspective of paleontology, the collections made by some of the early explorers were at that time of immense scientific importance. It was, after all, only half a century since the publication of Charles Darwin's controversial *On the Origin of Species*, which outlined his theory of evolution by natural selection, and at a time when drifting continents and the great southern supercontinent of Gondwana were fanciful concepts.

In terms of the birth of the great scientific ideas that these specimens were to inspire, the rocks and fossils that those heroic men collected are some small justification for the hardships they endured and, in some cases, for the ultimate sacrifices of their lives that they made.

2

Of Heroes, Rocks, and Fossils

It's not getting to the pole that counts. It's what you learn of scientific value on the way. Plus the fact that you get there and back without being killed.
—Robert Byrd

Before getting to the story of my fossil-hunting expeditions, it is interesting to understand how important the early Antarctic explorers regarded any fossil finds, in some cases dragging the rock and fossil specimens with them to their deaths, as in Scott's last expedition.

The very first fossils found from the Antarctic region were pieces of fossilized wood from Seymour Island collected by Captain Carl Larsen, a Norwegian who sailed south on the whaler Jason and landed on the Antarctic Peninsula over the austral summer of 1892-93. Afterwards he sailed on to reach 68° south, which was at the time the furthest southern latitude attained by any sailing trip, before having to turn back. The *Jason*, after having been at sea for several months and running short of supplies, met up with another whaling ship, the *Balaena*, on 28 December 1892. Aagaard recorded that some of the crewmen traded fossils for tobacco, quoting them as saying, "What were fossils good for when you had Navy cut and juicy quids?"

Larsen was a shrewd man who exchanged some of his brandy for pipe tobacco and kept hold of his fossils, which he would later give to the University of Oslo. These were formally described soon after as the first fossils from Antarctica.

In 1902 Swedish geologist Otto Nordenskjöld made the first major sledge journey in Antarctica, collecting many fine fossils and geological specimens from the western islands off the Antarctic Peninsula. His group wintered over on Snow Hill Island. Their ship, the *Antarctic*, dropped some men off at Hope Bay and shortly after was crushed in the pack ice. The men then discovered Jurassic plant fossils at Hope Bay. The crew of the ship spent a winter stranded on Paulet Island while Nordenskjöld and his men endured a second savage winter on Snow Hill Island, waiting patiently until they were finally rescued by an Argentinean ship in November 1903. During the long period he spent on the Antarctic Peninsula, Nordenskjöld also visited Seymour Island where he discovered the bones of giant penguins nearly 1.5 meters tall, a great diversity of shells and other marine fauna and well-preserved fossil leaves which led him to believe that, at one time in the past, Antarctica had been a much warmer place.

The fossil remains of archaeocyathids, a type of ancient, cup-shaped, sponge-like animal, were collected from the Antarctic region by the Scottish Antarctic Expedition of 1901-02; they were found by dredging up rock samples from the floor of the Weddell Sea. These were formally described by Dr. W.T. Gordon in 1920, who recognized that they were from the Cambrian Period, around 520 million years old, thus establishing the oldest age then known for the continent of Antarctica.

Ernest Shackleton's 1907-09 Nimrod expedition, driven largely by his determination to be the first to reach the South Pole, also resulted in many fine geological samples being collected. Isolated boulders found in the moraine and pebbles transported by the moving ice of glaciers in the upper reaches of the Beardmore Glacier contained small fossil archaeocyathids. An excellent monograph on the geological study of Shackleton's expedition specimens was published in 1914 by two of the participants, Australia's eminent professor of geology Sir Edgeworth David and British geologist Raymond Priestley. Their book included a special section on the paleontology written by Griffith Taylor and E.J. Goddard, which demonstrated that parts of the Transantarctic Mountains must be of Cambrian age.

The first fossils of vertebrate animals (fishes, amphibians, reptiles, birds, and mammals) to be collected on the mainland of Antarctica were bones of 380 million-year-old fishes found in Devonian rocks at the Mackay Glacier near Granite Harbour. They were collected in the summer of 1911-12 by Frank Debenham, an Australian scientist who joined Scott's Terra Nova expedition. Debenham's samples were pieces of the fossiliferous Aztec Siltstone containing mostly fragmentary bones, teeth, and scales of primitive fishes. The specimens probably originated further inland at Mt. Suess and had been carried along as glacial moraine. The eminent British Museum paleontologist Sir Arthur Smith-Woodward published a paper in 1921 in the scientific reports of the Terra Nova expedition describing these fossils. He identified some eight different kinds of fossil fishes and named three new species.

The significance of this material lay in the fact that Smith-Woodward recognized what was essentially a fauna similar to those found in the "Old Red Sandstone" then well known throughout many parts of Scotland and Europe. This was the first recognition of such a fauna in the Southern Hemisphere, and enabled a date of Late Devonian age (around 370 million years old) to be assigned to the rocks. At this stage, though, the cosmopolitan nature of the fish fossils was not taken to mean anything implicit about the past positions of the continents in question, just that freshwater fishes were assumed to be widespread around the globe at that time.

The incredible significance of the study of Antarctic geology and the question of the age of the frozen continent is best revealed through the dramatic entries in Scott's last diary. In early 1912 Scott's party of five struggled back towards their base camp after reaching the South Pole on 17 January and despondently finding Amundsen's victory message that he'd beaten them there by a month. Struggling along, low on food, scarred and weather-beaten by the harsh climate, they still managed to stop at the Beardmore Glacier and collect some 16 kilos of rock and fossil samples. It was the only day on Scott's final arduous march homewards from the pole that they were to devote to "geologizing." Scott's entry for his diary on Thursday, 8 February 1912 exudes excitement as he recalls the discoveries they made:

> The moraine was obviously so interesting that when we had advanced some miles and got out of the wind, I decided to camp and spend the rest of the day geologizing. It has been extremely interesting . . . Altogether we had a

> most interesting afternoon, but the sun has just reached us, a little obscured by night haze. A lot could be written on the delight of setting foot on rock after fourteen weeks of snow and ice and nearly seven out of sight of all else. It is like going ashore on a sea voyage.

Any knowledge of the remote inland geology of the continent was considered of great importance to science in those times. Only nine days later things were getting harsher, and Scott's conscience turned to the subject of lugging the rocks back with them. Evans had wandered off into the snow that day and was found frostbitten with "a wild look in his eyes." He was practically unconscious by the time they got him into the tent and he died quietly at 12:30 A.M. the next morning. On 16 or 17 March (Scott was unsure of the exact date), they were feeling despondent about making it back alive. They off-loaded a lot of their equipment but, due to Wilson's urging, kept the geological specimens. He wrote again about the burden of the rocks:

> The cold is intense, –40° at midday. My companions are unendingly cheerful, but we are all on the verge of serious frostbites, and though we constantly talk of fetching through I don't think any one of us believes it in his heart.
>
> We are cold on the march now, and at all times except meals. Yesterday we had to lie up for a blizzard and today we move dreadfully slowly. We are at No. 14 pony camp, only two pony marches from One Ton Depot. We leave here our theodolite, a camera, and Oates' sleeping bags. Diaries, &c., and geological specimens carried at Wilson's special request, will be found with us on our sledge.

This last remark implied that Scott knew instinctively that they would not make it back alive. Still, they carried the rocks onwards and the specimens were later found with their frozen bodies at their last camp. The men were buried in their tent where their bodies will remain frozen in their eternal, youthful state, until the slow action of the glacial ice eventually moves them out to sea.

Today the rocks and fossils collected by Scott's party are housed in the collections of the Natural History Museum in London and at the Scott Polar Research Institute in Cambridge. However, the specimens were not collected in vain. In 1914 British paleontologist A.C. Seward described the fossil plant remains collected by Scott's party from the Beardmore Glacier region. Although Ernest Shackleton's expedition had also stopped at the Beardmore Glacier and collected various geological specimens in 1908, four years earlier than Scott's party, they had not found any plant fossils. As is so often the case in the fossil business, those who search the hardest ultimately find the best specimens.

Seward made an amazing discovery. He was able to identify two genera of plants in Scott's specimens, *Glossopteris* and *Vertebraria*, both well-known forms from countries now not close to Antarctica. *Glossopteris*, for example, was an extinct seed fern with distinctive leaf patterns. It was known at that time from India, Australia, and South Africa, but not from any of the Northern Hemisphere countries. Thus the first seeds of the concept of "Gondwana" as a giant southern supercontinent were inadvertently sown.

Since the 1950s many geological expeditions have explored parts of the Transantarctic Mountains and the Antarctic Peninsula, both of which contain rich exposures of fossil-bearing strata, and their discoveries have filled out many of the missing gaps in the fossil record of the continent. Antarctica is now revealing itself as being home to an almost continuous pageant of life, represented by fossils ranging from simple invertebrates and algae at the beginning of the Cambrian Period, 540 million years ago, through to recently discovered fossil whales and dolphins about three million years old. Shell assemblages, only tens of thousands of years old, were discovered in the McMurdo district during Scott's days.

In his 1968 book, *Men and Dinosaurs*, well-known American paleontologist Edwin H. Colbert posed the question: "Will dinosaurs some day be found on the Island Continent of Antarctica? That is a question of great importance, a tantalizing question that is at the back of the minds of many dinosaur hunters today."

It was not until the mid-1980s that the first dinosaur bones were found in Antarctica, on James Ross Island off western Antarctica, by a team of scientists from Argentina. These were the fragmentary remains of an armored dinosaur, an ankylosaur. During the following years other bones of plant-eating dinosaurs, such as hypsilophodontids, were uncovered from the same region of peninsular Antarctica. The most spectacular dinosaur find occurred in 1990 in the central part of the

Transantarctic Mountains. A large carnivorous dinosaur was discovered by geologist David Elliot of the Byrd Polar Research Center, Ohio, in Early Jurassic sandstones (around 190 million years old) near the top of Mt. Kirkpatrick, which rises to more than 4,000 meters, undoubtedly the highest fossil locality anywhere in Antarctica. When paleontologist Bill Hammer from Augustana College, Illinois excavated the bones in 1991 he found even more remains of other dinosaurs and fossil reptiles from the site.

The dinosaur was a predator like the well-known *Allosaurus* and was later christened *Cryolophosaurus ellioti* by Hammer and his colleague, Bill Hickerson. It is the most complete dinosaur skeleton so far recovered from Antarctica. *Cryolophosaurus* was about seven or eight meters long and sported unusual swept-back bony crests above its eyes, giving it a kind of weird Elvis Presley look. In 1996 I visited Augustana College and studied the original specimen. It's truly an amazing find and one that, when considered in the context of the unusual dinosaur faunas of Australia, opens up a whole range of interesting questions concerning how and where different families of dinosaurs originated, and when and how they may have migrated around the globe.

Until all the Transantarctic Mountains are climbed and searched, we can only guess at how many more priceless scientific treasures are still waiting to be discovered. Our 1991-92 expedition had a mission to map the geology of unknown regions and collect fossils that would fill out the bigger picture of what Antarctica's ancient environments were like in the Devonian period (355-408 million years ago).

The trouble with collecting so many fine specimens, and so much new geological data, is that it takes several years to prepare and study all the specimens and synthesize the new results with previously published information. As this book goes to press, we have since published a series of scientific papers based on both the new material we collected combined with the older collections, resulting in some five new genera of lobe-finned fishes being described, three new genera of sharks and one new placoderm, an extinct group of armored shark-like fish. Publication details are listed at the end of this book.

There still remain some undescribed forms of lungfishes, new placoderms, new types of acanthodians (an extinct group of spiny fishes), and a new ray-finned fish (the largest modern group of fishes which include forms like the trout and goldfish) that we are currently working on. The goal of our systematic work is to first describe the fossils we have, and thus identify what species are present in the Antarctic fauna, then make useful comparisons with similar faunas from nearby Gondwana countries. We can then use these observations to test existing hypotheses concerning the nature of plate tectonic reconstructions, or the refinement of the age of fossil-bearing deposits throughout Gondwana.

People often ask me how useful is it to study the long-dead fishes of Antarctica. How can it benefit society? Well, here's a simple example of how our work down in Antarctica has been of some use in recent years.

In late 1996 a team of paleontologists from South Africa invited me over to study their Middle Devonian fish faunas and to participate in a field trip with the intention of collecting more specimens. During that trip to the scenic Cedarberg Mountains, about 300 kilometers north of Cape Town, we discovered some new sites that were rich in Devonian fish fossils, although most sites yielded only fragmentary bits and pieces of the fishes. One of these sites, a remote spot high up in the Cedarberg Mountains, contained abundant fossilized shark's teeth. At the time I instantly recognized some of these as belonging to the same new species I had found in Antarctica! Here then, for the first time, was a major continental correlation between the Aztec Siltstone of Antarctica and the Klipbokkop and Adolphspoort formations of South Africa. This discovery not only reinforced and confirmed the age of the South African fish deposits, but supported the Gondwana connection between Antarctica and South Africa in Middle Devonian times.

In the search for new economic deposits of minerals, oil, or gas, the dating of rock layers using fossils is a valuable tool that can save the exploration companies a lot of expensive drilling, or even point the way to more likely prospects through the environmental information the fossils provide. Fossils therefore give important information that contributes to the overall picture of the geology of a larger region.

There is more to paleontology than just using fossils to estimate the ages of rocks or their past environmental settings. Let us look for a moment at Antarctica as a continent. It is the fifth largest continent on

Earth, much larger than Australia, and about 1.5 times the size of the United States of America, covering approximately thirteen million square kilometers. More than 90 percent of Antarctica has an ice sheet nearly four kilometers thick over it. This makes Antarctica the world's highest continent, with an average relief of about 2.7 kilometers above sea level. Most of the world's fresh water is locked up as ice in Antarctica and if this ice sheet ever thawed out entirely the seas of the world would rise approximately 60 meters. This fact alone is of ample significance to justify our endeavors in understanding the prehistory of Antarctica in order to decipher the trends of its ongoing climatic evolution.

The Antarctic Treaty, which was ratified in 1961 and signed by a dozen participating countries, precludes the economic mining or exploitation of resources from Antarctica. Still, the study of Antarctic geology and paleontology is vital to understanding the big picture of Gondwana during the past times when no political or physical boundaries existed between Antarctica, Australia, Africa, South America, India, and much of the Middle East as we know it today (incidentally, the richest oil-producing countries in the world). If you drive a car and need petrol, you need fossils. Most petroleum deposits around the world have been found using the microscopic fossils within the layers of rocks the drill brings up as a guide to drilling depth. So, even esoteric research like the study of ancient fishes that once swam around Antarctica nearly 400 million years ago may prove to be of benefit to society's needs somewhere down the line. It's that simple.

Moreover, all economic considerations aside, I feel that we need to understand Antarctica as an evolving continent, from the study of its prehistoric environments and extinct animal and plant life, because it is all part of the global knowledge that contributes to our present understanding of the world as a whole living, evolving system. Without knowing Antarctica's full story, how can we possibly interpret our fluctuating climates and global weather patterns? How can we possibly predict global warming and its impact on future societies until we know how much of the world's weather is reliant on factors emanating from Antarctica?

Fossils are the prime way of deciphering the past climatic regimes of the Earth, giving a yardstick by which to measure the nature of climatic variation through time. The methods of doing this involve studying the climatic constraints of living species and making inferences about how closely related fossil species would have lived. In addition, geochemical analysis of the ratios of oxygen isotopes, formed in the shells of marine organisms, can be used to provide accurate data on the range of temperatures of seas during prehistoric times.

Large biogenically produced structures, such as reefs, seem to have always been climatically constrained due to the high-energy requirements of their existence. Therefore the presence of such large-scale fossil reefs is generally indicative of that part of the world having once been within a tropical zone. So, to make a long story short, these are just some of the many ways that fossils contribute towards reconstructing past climates. In order to fill out this picture more data are required from many parts of the world so that scientists can fine-tune the existing model. Any information on the fossil assemblages of Antarctica is therefore significant in contributing valuable information towards the big picture of the Earth's climatic evolution.

Other areas of Antarctic science are extremely important to our global community. The hole in the ozone layer was discovered by British scientists in Antarctica working on atmospheric physics. Chemists working in remote parts of the polar plateau of Antarctica can measure the subtle levels of global pollution by analyzing minute amounts of metals like arsenic or mercury from "pristine" ice core samples, and give us a world measure of pollution increase on a year to year basis.

Zoologists studying nematode worms locked in the sea ice for tens of years have discovered that they have the ability to come back to life after being frozen solid. Implications from such discoveries for effective cryogenics, or the possibility of being able to freeze people for long periods of space travel in the distant future, are nothing short of mind-boggling. And finally, let us not overlook the vast biogenic productivity of the shelf seas around Antarctica as another possible vital food source for the world, a subject to which I shall return later in this book.

Today Antarctica is the world's laboratory, a place of peace and cooperation where many different disciplines of valuable scientific research are conducted by many different nations. It is also opening up to the increasing demands of tourism, for one of its major and most important resources will always be its unique beauty, its primal appeal to the sense of adventure in all of us.

Yet the only way anyone gets to go to Antarctica and travel extensively inland to its most remote regions is generally through participation in a scientific research program. Even if you can get yourself involved in such an expedition, with full approval and funding to support your venture, as I did, one must still undertake all the necessary preparations and training so as not to succumb to its many dangers.

And this I did, at Tekapo, in the mountainous center of the South Island of New Zealand, in 1988.

3
Survival Training: Tekapo, New Zealand

In all things, success depends upon previous preparation, and without such preparation there is sure to be failure.
—Confucius

Antarctica is a deadly continent, a place where under normal conditions, death would be the usual human state of existence! Despite all the precautions taken each year fatal accidents keep occurring there. It's a simple fact of life in Antarctica that violent weather can close in around your field party in minutes, whiteout conditions can envelop your aircraft almost without warning, or the ground you are traveling over can suddenly become riddled with deadly crevasses which were previously hidden from view by a light fall of snow. My story begins with a brief insight into the nature of preparation for survival in Antarctica.

Being part of an official New Zealand field expedition it was compulsory for me to do my basic training in the lower half of the South Island. Here, icy mountaintops provided a good environment in which to experience near-polar conditions. However, before even getting to go on the training camp one must satisfy stringent medical requirements. Only expeditioners in the best of health are allowed to go on a remote deep field expedition, not only for their own health, but also because the rest of their field party will be depending on them. Extensive medical and dental checks are required. I was sent a special medical assessment form by the New Zealand Antarctic Research Program (NZARP) to take to my doctor so he could check out everything on the list.

The medical took nearly an hour, and included taking blood samples to be analyzed for HIV, hepatitis, and other diseases. The dental checks are also rigorous and resulted in my having all my teeth radiographed to check whether any of the older fillings might be at risk in the extreme cold climate. I had heard stories from old expeditioners about teeth problems due to the cold. Noel Barber's book about the 1955-57 Transantarctic Expedition called *The White Desert* touches on this subject:

> If you were working outside and talked too much, the cold contracted the fillings that then dropped out. I lost one that way, fortunately not one that caused any pain. At McMurdo the dentist was the busiest man in the camp, for he was the only one. I never dared to ask him what would happen if he ever got a toothache himself.

Once all the tests had been done, I knew for the first time in my life that I really was healthy. Finally, after I had sent all the medical reports back to the NZARP, I was allowed to attend the weeklong survival training camp at Tekapo, in the South Island of New Zealand.

In addition to Margaret Bradshaw, whom I'd met earlier in Christchurch, my field party consisted of one other scientist, Dr. Fraka Harmsen from the California State University in Fresno. Fraka was a Dutch-born girl who grew up in Nelson, on the South Island of New Zealand, and completed her doctoral studies on carbonate sedimentary rocks at Victoria University in Wellington. She was then appointed as a lecturer in sedimentology in the Geology Department of California State University in Fresno, USA. She was an expert in the study of sedimentary rocks and how to interpret the ancient environments in which they formed. She was a small, frail-looking woman with shoulder-length curly blond hair and an almost perpetual grin. She didn't look anything like someone who you imagine would be comfortable in the frozen wilds of Antarctica.

In August 1988 we arrived at a small military camp situated near Lake Tekapo after a scenic bus ride down from Christchurch. It was situated in a truly beautiful part of the world, set amongst high snow-capped mountains and valleys laden with deep green conifer trees and ferns, a land of kiwis and keas (large, brownish-green New Zealand parrots). Clear flowing aqua-blue rivers twisted and rolled over rocks forming white frothing rapids. The air was invigoratingly crisp and clean.

The base had that sort of "school camp" feel about it. Being in a military environment meant that rules were there to be obeyed, so times for meals and attending courses had to be strictly adhered to. No sleeping in late here. I was keen to learn anything and everything to ensure my safety and comfort in Antarctica. I had always lived by the motto that "any fool can be uncomfortable in the bush," an old adage which meant that if you prepare yourself well you can spend your time camping out in the desolate Australian outback and be perfectly comfortable.

Training involved learning many varied skills essential for survival: serious first aid, including how to give injections and perform your own emergency dental jobs; mastering the various radio communications devices; delving into the mechanical anatomy of two-stroke 500 cc Skidoo engines; how to use and service the three different kinds of chemical fire extinguishers used on the base; how to use and maintain the primus cooking equipment; how to erect a polar tent in a blizzard; the fundamentals of basic mountain climbing, rope work, and crevasse rescue skills; and how to cope with psychological challenges, like the isolation and problems of homesickness. Even personal hygiene was drilled into us, emphasising how important it was simply to brush your teeth daily to prevent gum disease, and to keep your feet dry and warm at all times, so as to prevent the development of "immersion foot," a sort of Antarctic version of tropical foot sores.

We were flat out all day for the first four days of solid courses and, at nights, after high cholesterol doses of good greasy tucker, we were subjected to rambling tales of hoary adventure by old crusty expeditioners. Sometimes we were made to watch gruesome medical slides of mountaineering accidents involving serious frostbites of frozen limbs, highlighting the kinds of amputations necessary once gangrene gets hold of you after the frostbite. These were explicit, grisly reminders of how important it was to always adjust your level of comfort by fine-tuning your layers of clothing and to keep a close eye on your mates at all times. It's common for people to develop the first signs of frostbite without being able to feel it themselves, as may occur on the nose or ears when they are totally preoccupied with serious work, or sledging on long journeys.

"Frostbite" is defined as a condition in which the flesh is frozen solid. It most commonly affects one's fingers, toes or the protruding parts of the face, like the nose or ears. If the frozen tissue is rapidly thawed in hot water, tissue can sometimes be saved. Shallow freezing of the exposed tissue is called "frostnip." This strikes nearly every Antarctic deep field expeditionary at some stage or another, yours truly included. Although painful, it usually causes no lasting damage once the affected region has thawed out.

The process of thawing out frostnipped flesh entails slowly warming up the afflicted parts. In my case it was often my toes, as early on in the field trip in 1991 I was not used to the extremely cold temperatures encountered in late October. At one time when we were out on a shakedown trip I felt the cold affecting my toes but couldn't be bothered having to stop the training to go inside a tent and change my boots. A short while later I found my toes going numb. Eventually I realized that I couldn't feel anything in my toes, so had to go inside the tent, warm up and change my shoes and socks. As the blood flowed back into those toes on warming it was excruciatingly painful.

The more serious condition, hypothermia, is when the body's core temperature, usually at 99°F, drops below critical levels. It is a far more serious condition and must be treated at once. As the body's core temperature drops to 86-90°F, unconsciousness occurs and death will follow if the core temperature drops below 77°F, as it is at this temperature the heart stops beating. The first serious signs of hypothermia include disorientation, wild erratic behavior, shivering, and fits and, finally, the victim goes into a torpor drifting eventually into unconsciousness. These days it is very rare for anyone working in Antarctica to be affected by hypothermia, as everyone is warned about it and made to take adequate clothing and preparation for outside field work. However, if by accident someone falls into icy water or down a crevasse and is only rescued after rapid hypothermia has set in, the person is best treated by rapid immersion in a hot bath of water at 108°F.

One evening we had to watch an amusing Canadian film about treating hypothermia. It must have been the corniest film I've ever seen, but it did get a few important points across and certainly made us all laugh. It presented the scenario of an absolute idiot wandering out in the snow by himself, getting lost, and subsequently showing the first signs of hypothermia. He starts wildly throwing off his clothes and runs around half naked in the snow until finally he falls asleep, exhausted, under a light snowfall. His mates find him a few hours later and he is in a mighty serious condition. The first thing they do is put up a small tent and then undress him and get him inside a sleeping bag with the two of them naked, so that their body warmth can slowly reheat him. This of course drew numerous snide remarks from the group, although it was actually the recommended life-saving procedure. The odd sexist remarks came out of the crowd about how someone hoped some good-looking lady in the group was going to get hypothermia so he could volunteer to "thaw her out." Despite the humor, the reality of frostbite and hypothermia hangs over all who venture on long outdoor travels in Antarctica.

Medical problems can get very serious in the Antarctic. In 1961 a Russian medic on a remote station actually took out his own appendix using two assistants to hold the mirrors and forceps. It was a matter of life or death, and he survived. The instruments he used are now on display in a medical museum in Moscow.

One of my favorite training courses that week was called "helicopter familiarization." This actually involved learning the correct procedure for getting on and off a chopper (without losing your head) and how to cope with "peculiar flying conditions" or other emergency situations that might arise while traveling in a chopper.

The latter was demonstrated to us when the pilot took us up very high then plummeted down in a spiral spin, making us all go white and dizzy. I lost my confidence in the crew then when the co-pilot threw up in the cabin of the chopper, a New Zealand Air Force UN-1N "Huey." The pilot luckily was an extremely competent fellow and managed to get us safely on the ground a few minutes later.

The final part of our training was to put everything we had learnt that week to the test up in the snowfields. When the chopper lifted us up to the snowy peaks, it hovered just above the snow and we had to jump out and unload our gear. The first thing I noticed was the wind and cold exacerbated by the chopper's whirling blades blasting the freezing air down on us. We managed to quickly off-load our gear from the chopper and get our tent up. After setting out our sleeping mats and bags we then adjourned to the snow to make a snow cave: a sort of igloo that you can quickly make to live in if your tent gets destroyed.

To make a snow cave the easy way, you just pile up your bags and gear into a mound and then heap loads of snow on top and flatten it down with a shovel. Tunneling in and pulling out the gear leave you with "Chateau Frigidaire," a makeshift snow dome. By scraping out more room at the base you could eventually end up with a comfortable shelter for the night out of the wind, with comfy terraced beds cut into the snow, little tables, or whatever else took your fancy (perhaps an ensuite). A person can be quite warm inside such a shelter as the air is still and heats up from your body warmth and the flames from the primus stove. If it gets too warm though, it starts to be a problem as the walls melt a little and your things get sodden. As long as you sleep inside your bags on a ground sheet, the temperature is really quite comfortable.

Many of the early expeditioners have testified to this, perhaps the most prominent amongst them being Victor Campbell's Northern Party of Scott's Terra Nova expedition. In 1912 six men lived through winter in an ice cave only a few meters long by a few meters wide by two meters high. They subsisted daily on the meat from penguins and seals and the warmth radiating out from their small blubber stove. They survived this way for six months, but not without some consequences of stomach illnesses. Nonetheless, their endurance and careful planning saved their lives.

Our day up on the mountaintop went smoothly. We were able to practice erecting our polar tents, setting up radio communications, firing up the primus and cooking a simple meal (something dehydrated that just needed water). We all slept well inside our down bags, and in the morning packed up camp and waited around to be picked up by the helicopters.

After the week at Tekapo had finished we all felt that we had learned and practiced the necessary survival skills for working in Antarctica, and were well aware that we would all be put through yet another survival training stint shortly after we arrived in Antarctica. It wasn't just good enough to learn these skills. We had to know them well and be able to perform every one of them in real Antarctic conditions.

I flew back to my home in Hobart, Tasmania, and now had only four months to wait until my first trip to the frozen continent would begin. So far I had passed the medical with flying colors. I had survived the survival training camp and was now ready for the next challenge, which would be the real thing, Antarctica.

For the remaining months before my trip I read avidly about Antarctica and the early exploits of the scientific expeditions. On reading about Sir Vivian Fuchs and Sir Edmund Hillary's heroic trans-Antarctic crossing of the continent using tractors in 1957-58, I couldn't help thinking about the British tabloid headlines of the day when the expedition was departing: "Sir Vivian Fuchs Off to Antarctica."

Yep, I thought to myself. Pretty soon I'd be doing the same thing.

4
Arrival in Antarctica

Rising steeply from the ocean in a stupendous mountain range, peak above peak, enveloped in a perpetual snow, and clustered together in countless groups resembling a vast mass of crystallization, which as the sun's rays were reflected on it, exhibited a scene of such unequalled magnificence and splendor as would baffle all power of language to portray or give the faintest conception of.

—Robert McCormick

Just imagine being the first person to lay eyes on the Transantarctic Mountains, a towering range of snow-capped mountains that seem to stretch forever skywards as jagged, threatening peaks. These words, from Robert McCormick, surgeon on board Captain James Ross' ship, the *Erebus*, demonstrate the overpowering wonder afforded by this awesome view. On arriving in Antarctica for the first time, I think it fair to say that everyone feels a little of this emotional power in their hearts. They will always remember that moment.

The first emotional high point of any such trip is saying goodbye to one's family. My wife, Donna, and I had three young children at that time: Sarah, aged 5, Peter, aged 3, and Madeleine, only 8 months old. It was going to be hard for me to leave them, especially knowing that Antarctica has its own brands of danger. Amidst teary goodbyes I departed the cold island of Tasmania on 9 December for yet another cold island, New Zealand, en route to the ultimately cold continent of Antarctica.

In New Zealand I stayed at the Windsor Hotel in Christchurch, a venue commonly used by Antarctic travelers, so it didn't take long before I'd met up with some other expeditioners also waiting to be called to the airport for the big trip down south. My other expedition members were already down in Antarctica in a remote location in the central Transantarctic Mountains, so I wouldn't be meeting up with them until some time after I'd arrived at Scott Base.

On the morning of Saturday, 12 December 1988, after waiting for two days of delayed flights, I was called to be ready for departure. All of us scheduled for that flight ate a hearty breakfast that morning, then put on our polar underwear, woolen fleecy trousers and shirt, and carried our packed kitbags to the front of the hotel. It was a pleasant summer morning and we felt a little awkward all kitted up in our polar clothing as we loitered around in the warm sunshine awaiting our lift to the airport. At 10:00 A.M. we were met by the New Zealand Antarctic Division representative and shuttled off to the US Naval Air Base located at Christchurch international airport.

Each person is allowed a maximum of 75 pounds of luggage, which is mostly the standard Antarctic clothing and boots, with a few personal items brought from home such as cameras, notes, books, and so on. As we lined up to present our bags for the weigh-in, a black Labrador sniffer dog gave each of us a serious drug inspection. It wasn't too bad—at least the dog wore a rubber glove. Then we were shown to the "departure lounge" to await our flight. We were on a full flight containing 38 passengers and nine crew on a C-130 Hercules equipped with ski landing gear. The flight was further delayed so we were allowed one hour for lunch at the base cafeteria. Luckily for us, one of our friends was with us, an American geologist named Noel Potter, so he was able to pay for our food with greenbacks, as only US currency was accepted there. I scoffed down a rather bland cheeseburger, some fries, and a Pepsi, and relished the thought that this greasy fast food would probably be my last such meal for several months. Finally, at about 2:00 P.M. we boarded the VXE-6 Squadron Hercules, now fully clad in our Antarctic clothing, complete with heavy survival jacket. The latter is a safety requirement in case the plane has to come down in some remote icy location.

The birth of VXE-6 Squadron came about from the Antarctic Development Squadron Six (VX-6) whose first flight to Antarctica was on 19 December 1955, when several planes set off from Wigwam RNZAF base near Christchurch bound for Antarctica at the start of the American Operation "Deep Freeze." That day several of the planes had to turn back to New Zealand as headwinds caused them to use up more fuel than they could carry to safely get them all the way down to Antarctica. The rest of the squadron continued and the first to land was Lt Commander Joseph Entrains Neptune. His comment on the journey was: "It's the most miserable flight I have ever made."

A year later VX-6 Squadron under the command of William "Trigger" Hawkes and Gus Shinn landed an R5D named *Que Sera Sera* on the South Pole. The crew became the first men to step foot on the pole since the 1912 parties of Amundsen and Scott. In January 1969 the squadron was re-designated VXE-6 Squadron. To date, the squadron has carried over 195,000 passengers to and from the Antarctic, carted more than 240 million pounds of dry cargo to Antarctica, and some 10 million gallons of fuel to sites down there. Using a variety of aircraft ranging from the P2V2 Neptune, UC-1 Otter, C-130 Hercules, C-140 Starlifter, R7D Super Constellation, the C-5 Galaxy, helicopters like the LH-34 and HUS-1A, and more recently the HH-1N Huey, they have had a remarkable safety record and have been awarded several honors for their impressive record at flying over the world's most dangerous continent.

On entering the plane, a US navy man at the door handed each of us lunch in a cardboard box and a set of earplugs. On looking inside the box, we pondered momentarily whether to eat our earplugs and stuff the sandwiches in our ears. We were soon crammed in to fit every available space, sitting there like rows of heavily insulated sardines. The narrow canvas seats were slung across a crude metal framework facing the center of the plane, and there were few windows to look out of. The flight was rather boring, the ear plugs and noisy engines precluded any kind of conversation with the other passengers, so I spent most of the flight reading my book (a cheery little novel called *Misery*, by Stephen King), with intervals of stretching my legs walking around the plane. Inside the box everything was wrapped in plastic at least once. We had a cheese roll with bright orange bland cheese, some cookies, a chocolate bar of some unknown, unclassifiable species, a cold chicken drumstick sealed in clear plastic (looking somewhat like forensic evidence at a court case), and an apple. Most of us ate the parts we liked (the chocolate bar), kept the apple for later and dumped the rest on exiting the plane.

As we approached Antarctica six hours later, someone informed me that we had passed the point of "no return" when more than half the fuel has been used. Several times each season a plane may be on its way to Antarctica and find out from the radio that the weather at the base has suddenly turned very bad. In this case, if there's enough fuel, it simply turns around and flies all the way back to Christchurch. These "boomerang" flights are reasonably frequent at the beginning and end of each summer field season when the weather is quite changeable. In our case the weather was fine so there was no turning back. Although I felt happy that we were going all the way, it was still a little scary for us Antarctic virgins who thought about the possibility that if the weather did turn bad at the last moment we would still attempt a landing. It had all been explained to me before, and I had confidence in the pilots, who were very experienced. The weather was forecast as being fine, so we had no cause for real concern.

Still, thoughts of the New Zealand DC-10 disaster of November 1979 lurked at the back of my mind during that flight. On that occasion a plane destined as a tourist flight over Antarctica had flown too low and crashed into Mt. Erebus, killing all 257 people on board. The wreckage still remains on the slopes of the volcano that overlooks Scott and McMurdo bases. Only recently have tourist flights over Antarctica resumed.

About three-quarters the way through the flight we were told by the cabin crew that we could come up to the cockpit one at a time to see the view. When I got up there I could see why people lined up for ages. What an amazing perspective on the world! The whole of northern Victoria Land stretched out below me, filling my whole field of view below the horizon with jagged snow-topped mountains, crevasse-ridden glaciers which flowed out to sea as prominent ice tongues, and vast white never-ending plateaus of snow—a clear blue sky above and nearly all white below. I looked hard to see any rock exposures but very few were apparent, only some craggy little bluffs poking out under the thick blanket of snow. It seemed like an impossible task to search for fossils down there, but I was reminded that we were still a long distance away from the central Transantarctic Mountains where our fish fossil sites were located.

The plane made a gentle touchdown at about 10:30 P.M. after missing the landing strip on the first try and having to effect a full 360° turnaround. It seemed like it was forever gliding along the ice runway before it finally stopped. On leaving the aircraft I saw the towering volcano, Mt. Erebus, looking absolutely humungous, peering down at me between the few wispy clouds floating around its smoking summit. As blasts of cold air raced around my face, scattering my icy breath, the fact that at 10:30 P.M. at night the sun was belting down on us and it was around –4°F, created an atmosphere quite alien, yet powerfully serene. It was without doubt incomparable to any place I'd ever been to before.

Dave Carrera, officer in charge of Scott Base that year, was waiting to greet us. Scott Base and McMurdo Base run one hour behind New Zealand time, so we reset our watches and drove along a well-made graded ice road to arrive at Scott Base around 10:00 P.M. Scott Base somewhat resembles a series of large green boxes up on stilts, joined together with corridors and walkways. It has a few large hangars and equipment storage sheds outside. Inside its large metal refrigerator-style doors it is surprisingly warm.

Scott Base was first established for New Zealand's participation in the International Geophysical Year (or IGY) of 1957, and to provide a support station for the Commonwealth Transantarctic Expedition. It was designed and prefabricated in New Zealand, then erected under the supervision of Frank Ponder with an army team in January and February of 1957. Situated at 77°51′03″ south, 166°45′45″ east, it sits on one of the lava tongues of Mt. Erebus on Ross Island. The first team that wintered over that year comprised 23 men under the leadership of Sir Edmund Hillary. Since then men and women have wintered over there every year, although the base has been much expanded and modernized since those early days. The original parts of the base are still used as overflow accommodation during the busy summer season.

Inside the base we changed out of our polar gear into our civvies as the base is always kept at a constant cozy 64°F inside. The officer in charge then quickly briefed us about base procedures, safety, and protocol, and then we were allowed to go to the bar for a much-anticipated drink.

Lo and behold my amazement on walking into the Scott Base bar for the first time and seeing a room full of mostly young people wearing bathers, surf shorts, and bikinis, their noses streaked with luminescent green and pink zinc cream, all dancing vigorously to the blaring tones of the Beach Boys' "I Wish They All Could Be California Girls." One group was standing on top of the pool table dancing in unison. No old crusty veteran explorers in fur coats here!

Upon entering the bar and finding a space in the corner near the window, someone thrust a chilled can of Steinlager into my hand. I looked out the window to the Ross Sea ice sheet spread out before me. Twin volcanoes, White Island and Black Island, were poking out of the ice on the horizon. Brilliant sunlight streamed in through the windows. I could see a few seals lying around lazily on the sea ice somewhat resembling big fat slugs. The party contained many visiting Americans over from McMurdo Base, and all in all, about 40 or so beach-loving polar party animals.

From time to time groups of merry beach revelers would open the large refrigerator-style door, jump outside and go rolling around in the snow for a minute or so, then come running back inside to get warm and savor more bevies. I watched with amusement, thinking that maybe they'd been here too long or something. Little was I to realize how similar I would become to these people with the passing of time in Antarctica.

I stayed for a few hours that night chatting with Ian Paintin, a lanky Kiwi geophysicist and keen explosives expert who had come down on our flight. Many people came up to us and introduced themselves once they realized we had just arrived, so the social niceties continued for some time. I ended up going to bed around 1:30 A.M., still revved up by the excitement of finally having arrived in Antarctica after years of planning. I closed the wooden shutters over the window to block out the sunlight in the bunk area, and then slept deeply.

The next day I awoke at 9:00 A.M. It was Sunday morning. Not surprisingly, there was hardly anyone about because most were enjoying the one day of the week allowed for sleeping in. After a light breakfast in the mess I changed into my polar gear and went for a walk outside around the base to explore my new home for the next few days. It was snowing and quite windy. I kept close to the base on the road and stayed out for a few hours, returning for lunch at noon.

Lunch is always very good at Scott Base, as were all the meals I had there: usually a hot dish of something with several types of salad and cold meats, freshly baked bread, and several fruit drinks to choose from, plus whatever fresh fruit is on hand. After eating, everyone takes their plates and cutlery to the sink and pre-washes their dishes before loading them into the large dishwasher. This is where I first experienced being "zapped." The air is so dry in Antarctica that static electricity builds up very easily, and then discharges whenever you come into contact with almost anything, especially things metallic. After rubbing your woolen and nylon clothing on the plastic-backed chairs in the mess, a huge charge would build up, just waiting for you to approach the washing-up sink or even just touch someone else. Zap! After the first few times, it became habitual to make a fist and whack the sink before the sink whacked you. The same applied all around the base; so before grabbing a doorknob or whatever, one had to always discharge the electrical build-up by hitting something metallic.

After lunch, Ian Paintin and I headed over to "Mactown" (the endearing nickname of McMurdo Base) on the bus, a Ford 4WD shuttle bus that flits between the bases every hour. We met up with Noel Potter, who gave us the grand tour of Mactown, Antarctica's largest mainland human settlement.

McMurdo Base is a big, sprawling base, somewhat disorganized but fully functional, like most American cities I've been to. It was first established during the IGY in 1957. Today it is a well laid out military-style base complete with nearby Williams Air Field (aka "Willy Field") and a well-developed harbor area for the incoming icebreaker ships. At the peak of each summer season it swells to hold up to 800 people. In 1962 the base housed a small nuclear power plant. Troubles with the plant forced its decommission in 1972, and with its removal went some 11000 cubic meters of radioactive contaminated rock. Back in 1988 I remember seeing lots of old fuel drums and redundant rusting equipment piled up around the place, but these days, so I'm assured, most of the rubbish has been carted away and stricter regulations are enforced to prevent any disposal of man-made material. Instead, the icebreaker ships that come at the end of each summer season now cart all waste back home.

First we visited the aquarium where the fishes were studied. Large seawater tanks, some about two meters in diameter, were stocked with a variety of local fishes. About 120 species of fishes are known from Antarctic waters, a surprisingly high diversity which is nearly twice the number of species known from the cold Arctic seas. The Antarctic cod were about one meter long and apparently very good eating. Sashimi made from their cheeks is said to be a rare delicacy that few people ever got to try, me included. There were several other smaller spiky fishes in other tanks (*Zanclorhynchus* and *Notothenia*). I could also see large marine isopods (*Glyptonotus*), looking like giant marine slaters, and some flower-like sea lilies (crinoids, a group of echinoderms related to starfishes). I was later invited out to the diving hut where we could watch the camera relay from a small submersible robot filming the bottom of the sea floor just out from the base. I remember vividly how these seemingly lifeless starfish-like crinoids would occasionally skip merrily along the sandy bottom if something threatened them.

The fishes actually live in sub-freezing waters which are at around 29°F. The seawater stays colder than freezing point due to its dissolved salt content. The fishes themselves do not freeze solid because they have anti-freeze glycopeptides in their blood, a marvelous evolutionary adaptation that enables these few fishes to take advantage of the abundant food supply around the polar shelf seas. In this respect Antarctica as a continent actually has a richer biomass than Australia, according to Australian ecoscientist Tim Flannery, even though most of this biomass is obviously concentrated in its shelf seas.

That afternoon the weather was overcast and somewhat dismal, with light snow falling almost continually. We walked about a kilometer to Hut Point and visited Scott's first hut, erected in 1901 on his Discovery expedition. Today, like all the Ross Island historic huts, the Discovery hut stands as an on-site museum, a veritable shrine to the heroism of the early explorers. The key is held at Scott Base, and one must get permission to visit from the Officer in Charge (OIC) and sign the register of visitors. It is immaculately restored after many years of painstaking work by the Antarctic Heritage Trust, a group of dedicated historians and volunteers from New Zealand. David Harrowfield, one of the most eccentric yet knowledgeable fellows I've ever had the pleasure to meet, was down there working on the huts almost every season. He has also published a book on the heritage sites of the Ross Island region.

When they first started work on the historic huts of Ross Island in the late 1950s the old buildings were in various sad states of disrepair, completely filled to the roof with the snow and ice built up over several decades. This was carefully excavated away, with roofs, windows, and doors repaired to as near the original condition as possible. Every artifact was labeled, catalogued and placed back in its original position as best determined from the excellent photographic records made by the early expeditions. Even the meat was preserved in the "cool room" (how ironic). Sides of mutton and seal meat turned whitish-grey still hang from the meat hooks. It inspires a feeling of silent timelessness. One could almost believe that at any moment Scott and his men would come shuffling in, hang up their deerskin coats and start boiling a brew.

Ian and I later had a coffee in the Mactown mess and caught the shuttle bus back to Scott Base. That evening we once more returned to Mactown as we'd been invited to see an art exhibition, so twelve of us piled into the OIC's Landcruiser and headed off over the hill, driving US-style on the "other" side of the road. The exhibition was great, and included lots of sketches, paintings, jewellery, and sculptures done by American base staff, mostly those who had wintered over at the base. They served us wine and nibbles while we took in the art. At 9:00 P.M. a few of us left for the Officers' Club to try some American beers. After one can of Budweiser (yuck!!) and one of Millers (not bad!) I left to catch the shuttle bus back to Scott Base and head for bed.

The following day we had a meeting with the operations manager, John Alexander, about how our event would proceed, and when we could fit in our survival-training course. In the afternoon I took a walk around the base and then set to work organizing my basic gear for the survival-training course I would be going on over the next two days. This is a compulsory course for all expeditioners, making us practice the survival skills learned at Tekapo as well as learn new skills more specific to moving around safely in Antarctica.

The survival training team who led our course was late coming in from Vanda Station that evening. A few of us went over to the field operations base in Mactown to hear a training lecture, and then we practiced a bit of "prussocking." This technique entails attaching loops of rope to our main climbing ropes that can slide up and down, enabling you to edge up the rope slowly. We also learned some fancy knots and rope work, and then watched some slides showing the techniques in action. The next two days we would be out in the field trying out these techniques, and we were duly advised on what gear to take with us to be comfortable.

After returning to Scott Base that night I had a couple of beers in the bar with Rod Sewell, Andy Allibone, and Jane Forsyth, all of whom are geologists with many field seasons' experience working in remote parts of Antarctica. We chatted about the rocks, sledging journeys and the general hazards of working in Antarctica till quite late. The Scott Base bar has a large photograph of Robert Falcon Scott up on the walls, along with snowshoes, field gear, and other assorted historic memorabilia decorating its cozy surrounds.

The picture of Scott beaming down at you combined with the caramel glow from low angle sun reflecting off the Ross Sea ice sheet is enough to make you want to get out there and start man-hauling your way to the pole! I felt fired up but first had to wait just a few more days until the necessary survival-training course was completed. I was also anxious to hear any news from Margaret Bradshaw's party out in the Darwin Glacier region, as I was scheduled to join their group as soon as they were pulled in from the field. The last I had heard was that they had been delayed, so I would have to bide my time at Scott Base a little while longer.

The photograph of Scott above the bar somehow haunted me. I couldn't help thinking of what had happened to him and his colleagues as I headed off to bed, and whether or not any degree of careful planning could really avert the direst of unexpected dangers in remote field situations. Tomorrow, I thought, I'd probably find out the answer at survival school.

5
Antics on Ice

So here ends the entries in this diary with the first chapter of our History. The future is in the lap of the gods; I can think of nothing left undone to deserve success.
—Robert Falcon Scott

Scott wrote these words on the night before he set out for his tragic attempt to reach the South Pole. Nonetheless, he expressed the sentiment that everything possible had been done to ensure their success, all possible preparations had been meticulously carried out and the men were well trained for the tasks each had to fulfill. Despite all this, the trip ended in disaster, so the emphasis on training and careful planning is now paramount to any fieldwork in Antarctica. I reflected on this as we were about to undertake our survival-training course, an important part of the modern preparation for any successful Antarctic field trip.

I awoke late next morning and rushed about madly to get ready for survival school. At 8:30 A.M. we departed from Scott Base in the Hagglunds snowmobile with John Alexander and JR, our survival teacher. The Hagglunds is like a giant enclosed tractor that pulls a trailer full of people and their gear along on twin tank treads. We traveled to a spot about six kilometers away from base onto the edge of Ross Island, yet still a reasonably safe distance away from the start of the sea ice, denoted by the pressure ridges formed by the frozen sea ice thrust up near the island's edge.

That morning we practiced walking up snow slopes and "arresting" techniques. No, this is not how to catch criminals on the ice, but rather what to do if you unexpectedly slip and go into a fast downhill slide. We were first made to slide down a steep snow slope upside down (head first) then turn ourselves around and stop the slide. I found my first attempt at this to be quite tricky, but after a few tries it became easy, and was actually a lot of fun. The arresting techniques involved getting up speed sliding down the snow slope, then turning and digging in the point of the ice pick in the bank to stop. It can be a little awkward once you are moving fast, upside down, looking up at the sky with your head leading the way down.

We then moved on to the "Chalet" for lunch. It was a marvelous-looking building, akin to a delightful postcard picture of a Swiss Alpine villa. It was actually a mobile house on sleds that could be towed around by a tractor to different locations. It made a comfy shelter in which to have a break from the wind on survival training courses, or if skiing on a day trip out from Scott Base.

After lunch we returned to the snowfield to build ourselves an ice shelter for the night. In the event of losing your tent in a blizzard, or if your sledge with all the gear on it goes down a crevasse, an ice cave is the best form of shelter to get away from the raging winds. To demonstrate this point, each expeditioner must construct either an igloo, above ground, or a below ground ice cave, and sleep the night in it. My group consisted of Rick, from the US Navy media services, and Tony, a US field operations man. Rick and I decided to go for the igloo, but Tony wanted his own personal ice cave. It was bloody hard work cutting the large snow blocks with a saw and building it up in decreasing circles so that it wouldn't collapse in on itself. After about three hours we were finished, the first shelter completed out of all the groups!

We next fired up the primus stove for dinner, a simple yet adequate repast of dehydrated lasagna (just add hot water), tea, crackers, and some instant pudding, all supplied by the US survival school. Others, who knew better, had brought along steaks and scampi, served with chilled champagne and other assorted luxuries. They cooked their meal up on a crude barbecue made from half a 44-gallon drum with spindly iron legs welded onto it. It was permanently kept out at the survival school training area. Those wanting to use the barbie had to bring their own wood from the bases, as it's not an easy job finding anything to burn in Antarctica, apart from penguins and seals near the sea ice, and they are nowadays protected. In order to light the barbie we used "woofer wood"—a can of petrol was poured over the wood and it was set alight with a mighty woof!

We all huddled around the burning wood for warmth. A roaring fire set outside on the Antarctic landscape in the late evening low light is indeed a peculiar sight, one that I never saw again. Following dinner we shared some port and chatted awhile until about 10:30 p.m. when the cold winds picked up, forcing us all to go inside our cozy ice shelters for the night. I slept very well inside the igloo, although it was a bit cramped for two people, as my feet couldn't extend fully. We now realized with humble hearts why we were the first group to finish, because our igloo was much smaller than the others! Luckily for us Tony built himself an ice cave, with a large snow platform for his bed, so he slept soundly below ground in his icy crypt.

We were all woken quite unwillingly by our unnaturally cheery survival school instructors at 7:30 a.m. The sun was shining and it was not too windy, so we breakfasted alfresco style. This was a simple task, as we only had to boil the primus for water to make tea. We ate dry crackers, tubs of instant pudding, and oranges, a wonderful balance of dietary goodness created by the US Survival School culinary experts.

After packing up the equipment we moved to the glacier to prepare for learning safe "glacier crossing techniques." This involved first spending an hour learning how to rope up our harnesses and walk around with crampons attached to our boots. Crampons are metallic spiky things that loop around on the soles of your boots so as to grip into the ice when you walk. Then we walked slowly up onto a glacier probing with our ice picks to detect crevasses. On finding one we would test its stability, and then cross it roped up to a partner who would potentially catch you dangling on the end of your rope should the thin ice bridge over the crevasse suddenly give way. Although this didn't happen to me, I was still quite relieved when this part of the training had finished.

Next we had to practice prussocking out of a crevasse and using anchoring techniques for stopping a fall. The first exercise involved having your buddy, who was roped up to you, suddenly throw himself wildly off to one side and you had to drop down and anchor yourself firmly to the ground using the ice pick and stop the rope from dragging you along with him or her. After this we went to a small ice cliff and practiced in turn arresting our buddy after he or she leaped off the edge. This was essential practice for a potentially life-threatening disaster, as crevasses are an all too common occurrence in real Antarctic fieldwork, as I would later discover.

Next we roped up and went to an ice cliff for a lesson on "rappelling." Rappelling is just another word for "abseiling" or letting yourself drop down over a cliff while controlling your rate of descent by tensioning your rope. We each abseiled down a twenty-meter ice cliff, and then solemnly waited, watching the others take turns to descend. Others had the chance to descend into a crevasse on a rope ladder, and see first hand what it's like inside one of those virtually bottomless icy chasms. I peered down over the edge into one and could see how difficult it might be to rescue someone who had fallen down into a crevasse. Not only was it a hell of a long way down to the bottom, but also it was just solid, smooth ice walls on each side.

The survival-training instructors told us a few true stories of accidents that had happened near the base when people had strayed away from the flagged safe routes in order to take a short cut over the glacier back to base. They were very unpleasant stories without happy endings. However, rather than dwell on the negative, here is an amazing true story of a crevasse rescue that was successful, as told by Finne Ronne in his book, *Antarctic Conquest*, about the 1947-48 US Antarctic expedition. On finding physicist Harry Clichy-Peterson down a crevasse, they miraculously rescued him:

> While we tied a bowline loop in the end of a rope I called, "What sort of shape's he in?"
>
> "Can't tell. Seems all right, but he's wedged head downward. It'll be a struggle."
>
> And a struggle it was. On the surface we could only guess what Butson was doing from the quivering of the ropes, as if they were fish lines with fish hooked on the ends. Actually Peterson was so tightly wedged that Butson could not get the loop we lowered him around the victim's body. Butson therefore had to untie the knot, work the end around Peterson's torso, and tie the knot again.

> "All right," Butson shouted up, "now pull! Carefully!"
>
> Four men heaved on this jury rig. At about the seventh or eighth pull, like a tooth being plucked from its socket, Peterson's body came loose from the jaws of ice. Then up he came, foot by foot, for a hundred and ten feet . . . he survived imprisonment for nearly 12 hours, as his thick alpaca-lined suit had saved him from the cold.

Unexpectedly, a large emperor penguin came sliding in from miles away to watch us. It moved rapidly by coasting on its belly, its little feet pushing it along like a rear propeller engine. On seeing it we all rushed towards it to take photos, as it was the first penguin most of us had seen in the wild. It stayed around watching us curiously, naturally posing for us as we photographed it. Then it headed off again just as quickly as it had arrived, its curiosity about us obviously satisfied.

Emperor penguins live in a large colony at Cape Byrd, the place where Wilson, Bowers, and Cherry-Garrard marched to in the dead middle of winter in 1911 to collect some of their eggs. Although the journey was only about 70 kilometers, it was in the dead of night and under the most extreme weather conditions that Antarctica could throw at human beings: blizzards and seriously low temperatures. Cherry-Garrard's book, *The Worst Journey in the World*, is aptly titled, and covers the whole of Scott's last expedition. The idea at the time was that penguin embryos might show some interesting aspects of their evolution, linking birds with more primitive kinds of vertebrates. The emperor penguins hatch their young in the middle of winter, the safest time away from predators, but they must huddle up in large groups to protect themselves from the weather. The young chicks are tucked under the skin of the males, who brood them while the intrepid females go out to sea in search of food. After a while they change around and the females take care of raising the young. By this time the fathers may have gone for up to three months without feeding. Emperor penguins are the largest of all penguins, and may stand as much as one meter high and weigh up to 30 kilos. They are certainly dignified-looking birds; their colorful orange-rimmed necks and black heads make them quite attractive.

After a quick lunch we packed up our gear and were taken back to Scott Base in the snowmobile. Following dinner that evening I was reluctantly dragged along to the annual Observation Hill Race at Mactown. Observation Hill is the black volcanic cone overlooking McMurdo Base. It has a large wooden cross on top constructed in memory of the deaths of Scott and his men. I originally didn't intend on running in the event but decided to join in at the last moment, fearing that my lack of fitness before being thrown out into the field would not be harmed one bit by this sudden burst of unplanned sporting activity. I ended up running the race in 13 minutes, from the fire station to the top, ranking 30th out of 50 entrants. We then went to a barbecue organized by the Kiwis from Fischer Catering. Fifty-six Kiwis were employed that year to work in Mactown, but most spent as much of their free time as possible drinking New Zealand beers in the Scott Base bar. Those of us who ran the race were then given vouchers for four free cans of beer. It was worthwhile participating after all, I reflected later.

After the feast we went to the living quarters of one of the Kiwis working at Mactown for a birthday party, where I mingled with about twenty people crowded in a small room. We had some cake but had to leave early as the US regulations prohibited "parties" in the living quarters. The warden for that area gave us the stern message by to leave at once, so the party broke up and we all headed back to our respective quarters for the night.

It was almost Christmas, and I still hadn't been able to get out into the field. No news had come in from Margaret Bradshaw's group, mainly because they were waiting to be picked up by VXE-6 Squadron and could do absolutely nothing to speed up the process. They were at the mercy of the weather and the mechanical fitness of the Hercules aircraft. In times like this it would have been very frustrating for them. After nearly two months working in extreme conditions they wanted nothing more than to get back to base but had to stay put and wait.

I went to bed that night hoping that news of our expedition would come in the next day, anything that would get our mission moving so that I could start doing the job I'd been sent down to do: search for fossils and uncover the secrets of ancient Gondwana.

6
A Very Good Christmas

Lying in the sleeping bag that day I dozed off into the land of food once more. This time it was a confectioner's shop, decidedly grandiose and apparently opulent. From amidst the heap of other customers crowding the counter, I commanded the attention of no less than the proprietor himself, who courteously led me by a winding stairway to the roof where, he explained, his most sacred productions were stocked. There to my amazed gaze were exposed two long rows of gigantic cakes, each about four feet in diameter. Never before had I imagined anything so glorious—for a brief space at least I was in Heaven. My delight was increased to a still further degree when it was explained to me that they were no ordinary cakes but that each was fitted with a fuse to be lighted just before serving, when the whole of the ingredients would react chemically (after the manner of termite) completing the cooking and delivering a steaming hot article. In rapture I ordered one, feeling disappointed that I could not carry more, and hurried down to the counter to settle for it. I remember paying the money over, but my next consciousness was the realization, as I walked down the street, of having omitted to carry off the prize itself. With all haste I returned to the shop only to be disappointed and mocked by discovering the door shut and on it the placard "early closing."

—Douglas Mawson

On Christmas Eve 1912, when Douglas Mawson was starving, with only meager rations and a little dog meat left, he had this most amazing dream of rich, sumptuous food which he wrote down in graphic detail. It highlights the importance of food in the daily routine of the Antarctic explorer, and to much the same extent it has a great significance to the people who live and work around the bases. I was unfortunate on this, my first trip to Antarctica, to be delayed in getting out into the field. On the other hand, though, being stranded at Scott Base over Christmas did have its benefits . . . .

I spent the morning of Thursday, 22 December in the Scott Base library, a quirky little upstairs corner of the base where shelves are overflowing with books and reports of all kinds. It's a great place to hide away for a few hours when you desire solitude, although in the peak of the season it can be a fairly busy place. I read over my field notes, studied the maps of our sites and dug up some interesting papers about the regional geology. I then watched a video for a while in the little video room while waiting for my washing to finish in the laundry. The only things usually requiring washing regularly are socks and underpants, but if you hang around long enough on base you also need to occasionally wash your bedding. Clothes and sheets are dried in a special drying room situated near the heating units that keep the base warm, so things dry very quickly. It's a bit like a sauna in there.

After lunch I went over to Mactown and met Dr. Bert Rowell, a paleontologist from the University of Kansas. Bert is a veteran of many Antarctic deep field expeditions on which he collected fossils throughout the remotest parts of the Transantarctic Mountains and mountains in western Antarctica. His first trip down there was in the late 1960s.

We chatted about his work on early Cambrian limestones, some of which formed huge reef-like structures. These are the oldest known fossils from the continent of Antarctica, dated at around 530 million years old. The kinds of creatures he had been studying included trilobites (segmented arthropod animals having three separate divisions), ancient snail shells, the sponge-like archaeocyathids, and a host of enigmatic beasties that we paleontologists wisely call "*Problematica*." The Cambrian reefs extended right across Antarctica, but good outcrops that could be studied in detail were dotted spasmodically through the Transantarctic Mountains—unfortunately, with quite large distances between them. Bert had been to almost every major Cambrian locality in Antarctica from the Pensacola Mountains in the far west (near the base of the Antarctic Peninsula) through to northern Victoria Land.

Such research is of great significance to our understanding of the palaeogeography of Gondwana at that time. For example, we know from much of the paleontological evidence that Antarctica formed the central hub of Gondwana then, but from detailed studies of the trilobites and mollusks we can also present hypotheses about which continents or smaller crustal blocks (called "terranes" in geological lingo) might have been close to or in marginal contact with Gondwana. Controversy raged at the 1991 Gondwana meeting in Hobart because Professor Ian Dalziel, from the University of Texas in Austin, presented a quite radical new paleogeographic reconstruction of Gondwana in the Cambrian period suggesting that North America was then adjacent to Australia. Some of us questioned the timing of this placement because similar groups of trilobites, like those denominated by the genus *Redlichia*, were then known to be restricted to places like Australia and Antarctica, but not found anywhere in North America. Through careful discussion and sharing of data the situation was eventually resolved: the trilobite evidence suggested that Dalziel's continental arrangement may have existed earlier, in accordance with the other geological data that still supported a close relationship between the two ancient landmasses. Research of this kind is of vital importance to tracing geological trends that facilitate the search for economic mineral deposits that occur through Gondwana regions, or imply conditions suitable for suites of similar mineralization styles.

I had the pleasure of catching up again with Bert in Kansas in 1998 where he showed me some of the fine specimens of trilobites and other invertebrate fossils that he and his colleagues had collected on their many expeditions throughout Antarctica. I spent a few days there looking at his material and photographing specimens for my own reference. Bert told me that he had made his last trip to Antarctica in 1996, and that at the age of 68 he had decided to hang up his mukluks and finally call it a day.

On leaving Mactown that afternoon I came back to Scott Base and did a workout in the "gym," a small room off the temporary living quarters that housed a few weight machines, cycling machines, and barbells. After dinner I spent some time reading, then later I adjourned to the bar for a few drinks, as was the regular evening habit.

"Moses" Turnbull, one of the NZARP expeditioners, walked into the bar that night in full expedition regalia. He had just returned from an expedition at 4:00 P.M. that day and was leaving on a departure flight at 9:00 P.M. that night, so he shouted the bar. Beers flowed freely for some hours. The tradition is that on leaving for home every expeditioner will go into the bar and ring the bell announcing that they are shouting everyone free drinks. As drinks are so cheap at the Scott Base bar (in 1988 it was one NZ dollar for anything at all), this is not an expensive gesture and is one greatly appreciated by all.

Next morning during breakfast I was summoned to the radio room. It was the much-awaited call from Margaret Bradshaw's group down on the Darwin Glacier. Margaret seemed worried by the fact that they could not get pulled out on time and was concerned that my scheduled field trip to collect fish fossils might now be in danger of not proceeding according to plan.

The story goes like this. Two groups were working in the same area and were due to be pulled out together. On Wednesday a Hercules went in to pick up the first group and while doing a ski drag to test the landing site they opened up a huge crevasse. They found another possible landing site six kilometers away but as the weather was foul they would have to come back and do another reconnaissance flight to photograph the site and analyze it for safety reasons before attempting a landing; so it looked like it would be some time before either group could be pulled out.

There was an amusing article about all of this published in the Christmas Eve issue of the *Antarctic Times*, a newsletter published by Scott Base. Here it is reproduced faithfully to show the nature of field humor under times of duress:

The Stranded Scientists Club

A meeting of the Stranded Scientists Club was convened on 11/12/88 at the Darwin Mountains Acclimatization Society Clubrooms on the Darwin Glacier. Present were Ken Woolfe, John Henare, and Grant Gillespie.

Discussions at the meeting were as follows:

—use of a D9 bulldozer the only sure way of crevasse detection

—whether a Herc could airdrop enough pallets to bridge all the crevasses on the skiway

—levitation is the only reliable method of transport in Antarctica.

The subject of low membership numbers was raised. It was felt this was due to the present location of the meeting, and the dry nature of the cellar. To this end, an airdrop has been arranged. It is expected that this may also help ensnarl four new members from K221.

Finally it was decided to make a special one-time-only free bonus offer: a one way Herc trip to the Darwin Glacier for all new members before 1/1/89.

The Stranded Scientists Club and the Darwin Mountains Acclimatization Society wish all new and potential members a Merry Christmas.

I told Margaret over the radio that my priority was to get to the Boomerang Range if our trip was going to be shortened. I was told that I had to minimize the number of sites I wanted to visit. The longer the delay in pulling out her field party, the less time we would have to search for fish fossils in the Boomerang Range area. It would also mean that it was less likely that Margaret's group would want to go back out there, as they had been out in deep field for nearly two months at this point and were growing tired of just waiting around to be pulled out. It was then suggested to me by the operations manager that if Margaret's group could not be returned to base by the next week that I should go out somewhere else nearby and do some geological field work to keep myself busy.

This was a worrying thought as it sounded like he could sense that for some unknown reason they wouldn't be coming in soon and that it was unlikely I would be going out to the Boomerang Range. He suggested I have a look at the geology of the Dry Valleys region to see if there was anything of interest at that location because it was relatively easy to get me there on a chopper flight over the next couple of days.

I felt let down at having to wait longer for my intended mission to happen. I passed the rest of the morning studying published reports in the library to find out as much as I could about the geology around Vanda Station. The Dry Valleys had some exposures of Devonian sedimentary rocks of the right age for finding fish fossils, so I worked for the rest of the day on drafting a proposal for exploring some potentially interesting areas around these valleys.

The next day was Saturday, 24 December, Christmas Eve. In the morning Andrew Allibone and I once more pored over the geological maps of the Dry Valleys region and finally completed an outline for a brief field trip into the region:

> Proposed geological trip to Lake Victoria and Wright Valley Regions
>
> Field Party: Dr. J.A. Long (K221), Mr. A. Allibone (K047).
>
> Outliers of the Lower Beacon Supergroup (Taylor Group) occurring near Sponsors Peak (Upper Victoria Valley), and in the Olympus and Asgaard Ranges (near Vanda Station) range in age from early to Middle Devonian (McKelvey et al. 1977, Bradshaw 1981). Although trace fossils are known from the rocks, no vertebrate remains have been recovered, nor has any vertebrate paleontologist been in the field to examine these sequences. We propose a reconnaissance trip to search specifically for vertebrate fossils as recent work done on the fish fossils of the uppermost unit (Aztec Siltstone) has been successful in age dating and geological correlation of the formation (Young 1988). In addition any invertebrate fossils or trace fossils will be noted or collected if any new examples come to hand.

The plan was to put us in on 27 December at Lake Victoria, spend the 28th working the Sponsor's Bluff region and on the 29th for two of us, including myself, to walk back through the Dry Valleys to Vanda Station. Then I would have two days to explore the Devonian rocks within hiking distance of Lake Vanda, with a planned pull-out around 1 or 2 January, depending on how logistics were progressing to bring Margaret's field party back to Scott Base. They would then need a few days' rest and recovery, have to carry out a re-supply of food, mend and exchange equipment but then we would (theoretically at least) be able to be dropped directly into the Boomerang Range for about a week's intensive field work. All we needed was the luck for all the weather conditions to be right, and available aircraft to carry out the immediate task of getting Margaret's party back in again. Not much to ask, you might think, but in Antarctica, at the end of a busy summer season, lots can go wrong.

We used the term the “A factor” to describe the fact that no matter how well you plan something, for unpredictable reasons you will lose about one day in every four when working in the field in Antarctica. If it’s not bad weather, it’s a mechanical failure, a delay waiting for someone or something to arrive, a bad spell with your health, or something completely unexpected happening. Whatever, you must plan with a large margin of extra time to accommodate the A factor.

I had come all the way down to Antarctica from Tasmania to do this work, now shortened from an original plan of two weeks’ fieldwork to about one week, if I was very lucky. So, once more I found myself having to bide time, powerless to do anything to change the situation. Like many scenarios in Antarctica one just has to be patient, and wait and hope that things will turn out for the best.

The rest of that day we were kept busy with various base duties. Our first chore was to walk out onto the sea ice and remove flags that marked an unsafe region of the pressure ridges. These are the huge jagged sheets of sea ice that have been pushed skywards by the action of the ice shelf rubbing up against the land. Occasionally holes open up where the ice is very thin, so these risky areas are flagged off as a warning to those walking around outside. As we wandered between the jagged sheets of saline ice we watched the Weddell seals lazily sleeping near their holes. Despite their sluggish look, these seals are apparently very fast and lively underwater, using their teeth to scratch at places where the sea ice is thin to make exit holes. We took close-up photographs of them, being careful not to get close enough to disturb them. After lunch everyone was summoned to participate in a general cleanup around the outside of Scott Base for about an hour or so. We were each given a black plastic garbage bag and went off in groups to pick up every scrap of man-made rubbish we could find.

One of the things I picked up that day was a small copper disc with a hole in it. It had no visible writing on it as it seemed to have been stained with age and weather. Later I would make a chain for it and wear it round my neck. Eventually after about a year of wearing it continuously my body oils cleaned up the surface, and I noticed it had some writing on it. It read “Stewart, 1974” and I suspect it must have been one of the dog tags for the huskies. The huskies were for a long time an integral part of any Antarctic sledging journey, and affectionate soul mates for any winter-over crews. They had been removed from Scott Base the year before (1987), due to pressure from the environmental lobbies because a few seals had to be killed each year to feed them. It was also argued that they were not a natural part of the Antarctic ecosystem, so it was better not to have them there. Note, though, how this argument, of course, doesn't hold for us humans!

Later that afternoon we had dinner and a couple of drinks in the Scott Base bar, and at around 8:00 P.M. we all went over to Mactown for Christmas Eve celebrations. We started at the "Chalet," the building that acts as headquarters for the US National Science Foundation, where there was joyous carol singing, hot eggnog punch, and a spread of delectable Christmas cake treats. It gave many of the American expeditioners a warm fuzzy feeling of what being home on Christmas Eve should have been like, although for an Australian like me, the cold winds and snow were about as foreign a sight for this time of the year as spotting a kangaroo surfing.

Nonetheless, I was keen to enjoy my first "white Christmas," which I heard about in my youth through the Bing Crosby song. So, after mingling with the American scientists for a while we moved on to the gym where a big party was raging, complete with a four-piece rock band playing very loudly. The high volume of the music clearly compensated for the band's lack of musical talent, but overall I thought they sounded pretty good. Soon it grew too noisy for us, so we decided to head for the sleazy, dark atmosphere of the Acy Deucy Bar, one of seven watering holes that existed at that time on McMurdo Base. Strangely, unlike the splendid Antarctic vistas afforded through the windows of the Scott Base bar, the American bars were all dark and dingy, possibly trying to recreate an atmosphere of being in a regular bar back home in the States, so no windows could be found that gave away the harsh reality of the situation.

Just before twelve we found a midnight mass going on, and squeezed on in to join in the carol singing and generally partake of the Christmas good spirit. We then went back to the Acy Deucy, which was now absolutely jam-packed full of drunken Christmas revelers, mostly US navy people. I thought at that point that I'd lost my bag, which had my camera in it, so I went off retracing all my steps on the base, first to the Chalet, then to the gym, and so on. On my return to the Acy Deucy someone told me that it was right there under our table! I whipped out the camera and took some photos of the crowded bar scene at about 1:20 A.M., just before closing time. The photos came out dark and surreal, slightly out of focus, the bar appearing overly crowded with twisted faces, almost like part of a scene from a Hieronymus Bosch painting.

I then decided to call it a night and head for Scott Base. My New Zealand mates had somehow left at different stages of the night, finally leaving me there alone with a group of American scientists. Unfortunately there were no shuttle services running between the two bases at this late hour (and on a public holiday), so I decided to hike it back alone. I had on a shirt and light jacket, jeans, and boots, so headed out into the icy winds and light snow for the two-kilometer trek over the hill to Scott Base. I made it back around 2:15 A.M., much to the delight of some late-owl Kiwis who had watched me continually slip over on the icy road as I made the final stretch down the hill. I woke up next morning slightly bruised, fuzzy in the head, but happy in the fact that I was here to enjoy a Christmas Day on Scott Base.

I arose around 11:00 A.M. on Christmas Day, showered, and went into the bar where everyone on Scott Base had gathered. Some were already guzzling down alcoholic drinks, but I started conservatively with some lemonade before eventually succumbing to a glass of Australian champagne, "just to get into the Christmas cheer." All Christmas day drinks from the Scott Base bar were complimentary.

Our entertainment started at around 11:30 A.M. "Don the Magician" had come all the way, live from his rave review show in (you guessed it) Mactown, to perform miracles of sleight of hand and other Antarctic illusionary wonders to mystify and stupefy us, not that that was hard at the time. His illusion of sawing the seal in half was great, apart from all the gushing blood (not really, just joking!). After being dazzled by the morning's entertainment I went to the communications room for my pre-booked telephone call back home. I got through to home on a hazy line and had a few brief minutes to speak with my wife and kids to wish them all a Merry Christmas from the deep white south.

I felt very sad to hear their voices and was overcome by pangs of homesickness. I missed them all dearly, but there was nothing else to do but join in the Christmas spirit and make the most of the day.

At around 2:15 we all adjourned to the dining room for Christmas dinner. The room had been beautifully set with fresh flowers on the tables, crisp white linen, and shining silverware. The dinner was truly amazing. The chef, Kerry, had been preparing for some days and had made sure all the necessary ingredients were flown in fresh on the Herc flight the day before. I will here describe the lunch, as, just for the record, it was the best damn meal I've ever eaten south of 66° latitude!

First course was pâté de foi gras with avocado and mint dressing, followed by a second course of delicately cooked baby squid tubes stuffed with other mixed seafood including prawns and scallops and covered with an asparagus and tomato dressing. Appropriate New Zealand and Australian wines were served with each course, perfectly chilled of course (what else would you expect in Antarctica!). The third course was a parcel of mushrooms in cream cheese wrapped in filo pastry with a blackberry sauce. Finally the main course arrived: large servings of perfectly cooked turkey, baked ham, and fresh vegetables. After three hours of stuffing ourselves, we succumbed to the traditional dessert of plum pudding with brandy sauce, fresh strawberries with various fruits and cream. Brewed coffee was then served accompanied by a wide range of liqueurs and ports, and served with a range of chocolates.

Let me explain a little about this extravagant lunch, just in case you think that life is all too cushy down on Scott Base. Normally the only people present on the base at Christmas are the regular base staff, mostly the winter-over crew and administrators, cooks, cleaners, trades people, and some scientific staff who maintain the laboratories and take meteorological and atmospheric measurements. Field scientists and expeditioners are usually out working in remote locations by then and so often have a meager Christmas celebration with whatever they have left in their ration boxes on the day, plus whatever extra small goodies they might have stowed away for Christmas. The Scott Base Christmas dinner feast is therefore a great morale booster for the hard workers who run the base all year round and who look forward very much to the Christmas Day celebrations. It was also a way in which the NZARP administration shows their appreciation for the hard work done by the base staff in keeping the huge logistics machine of Scott Base running smoothly. I was very privileged to be on the base for that Christmas Day, but as the reports on the previous days have shown, it was purely by chance that I was there that day rather than out working in the field.

After having consumed one of the most sumptuous meals ever to be created in Antarctica, everyone shuffled back into the bar to await the coming of "Father Christmas" whom, we discovered, doesn't really live at the North Pole, contrary to popular belief, but has a condo at the South Pole. Everyone present on base had bought a small anonymous present and put it into the bag so we were all to receive some small token of another person's good will. In order to get our present we each had to go up in turn and sit on Santa's knee and tell him what a good little boy or girl we had been, much to the amusement of the wisecracking audience. My present was a deck of cards.

After this a few people stayed on drinking in the bar while others disappeared for a snooze, me included. I woke up, went up to the library for a while and read, then to bed around 10:30 P.M. It had been, cuisine-wise, without doubt the best Christmas dinner of my life. My childhood memories were of scorching hot Melbourne Christmas Days where my mother cooked up baked turkey and ham and every year served up the traditional hot plum pudding with old silver coins in it. In Antarctica it was a truly white Christmas, an experience I'll never forget.

I woke up around 11:00 A.M. on 26 December and spent the next hour writing up my notes and letters home; I then passed a few hours reading until lunchtime. I spent this day quietly preparing for the field trip the next day to Lake Vanda by getting maps photocopied and going once more over published scientific papers on the geology of the region. My hope was to try and search the Devonian rocks of the region to see if they contained any fish fossils and to keep an eye out for any other kinds of fossils.

No fish fossils had ever been found in the region, not that this fact bothered me much. Fossils are often found by those with an eye trained to recognize them, especially when they are preserved in an unusual manner. For example, in Victoria where I had collected from many Devonian age fossil fish localities, sometimes the bone would be completely weathered away and I would only recognize the fish fossils by the patterns made from their impressions in the rock. Later, when cast with black latex rubber, these insignificant looking holes in the rock would reveal stunning detail of the surfaces of different fossil bones.

All my hopes now rested on whether or not the Devonian sediments of the Dry Valleys region had any potential to yield vertebrate fossils. Scott Base and McMurdo Base are nestled on the flanks of the smoking cone of Mt. Erebus on Ross Island. The next day I would leave the comfort of Scott Base to venture for my first foray onto mainland Antarctica. We had a long day of fieldwork planned, so I decided to give the Scott Base bar a miss that night and headed to bed early.

7

Dry Valleys, Sand Dunes, and Rivers

Certainly we were in one of the strangest, weirdest, and most terrible of all the corners of the earth's globe. Of all existing lands it was infinitely the most ancient. The conviction grew upon us that this hideous upland must have been the fabled nightmare plateau of Leng . . . .

—H.P. Lovecraft

The big orange whoop whoop bird arrived at Scott Base around noon. We quickly loaded up our gear into the helo and were flown over to Vanda Station in the heart of the Dry Valleys. We soared low over the ice shelf, cruising the margin where the sea and ice met. Below us were countless small dots of life; there were seals and penguins safely resting on the ice, and in the deep blue, cold water were the black and white shapes of killer whales keenly patrolling the edge of the ice sheet for unwary prey.

As we headed up the Wright Valley my mental image of Antarctica rapidly changed for we had entered one of the most remarkable dry valley systems on the continent. Here was a large area of exposed clean rock and sand (some 1860 square miles), yet bounded on all sides by ice and snow. It was below the level of the polar plateau, like a sunken lost world, quite reminiscent of Lovecraft's words describing the "Mountains of Madness."

The Dry Valleys region is an area of the Transantarctic Mountains inland from Ross Island that remains ice and snow free all year round due to the fact that the land is rising at a faster rate than the glaciers can encroach on them. Also, the howling dry winds that roar through there in winter make it impossible for any build-up of snow and ice to get started. This region was first discovered by error on one of Scott's earlier explorations during the 1901-04 Discovery expedition. Having overshot the mark in selecting a passage through the Transantarctic Mountains on their return journey from the east Antarctic ice sheet, Scott and his party inadvertently stumbled into the Dry Valleys in December 1903. However, being low on food and equipped only for snow sledging, they could not take the exploration any further. In 1911 Scott sent Australian geologist T. Griffith Taylor to explore the edge of these valleys, but it wasn't until the 1950s that the main valley system centered on the Wright Valley was discovered and explored for the first time. During the International Geophysical Year in 1957 a number of intensive studies were initiated to study the natural environment and geology of the Dry Valleys. Both New Zealand and America have had small base presences there ever since.

We cruised up the wide, clear Wright Valley, the towering mountains of the Asgaard and Olympus Ranges forming its scalloped sides and looming way above our heads, even in the helo. It was a truly breathtaking sight as the rocky walls are variously colored with irregular lashings of black dolerite sills, abundant light grey granites, and thinner layers of brownish shales and buff-white sandstones topping many of the pointed snow-capped peaks. We landed at Vanda Station around 12:30 P.M. and we then spent two hours there having lunch and waiting for our helicopter to take us on to Lake Victoria. Our team consisted of Chris Rudge, from the New Zealand Science Department's public relations group, geology student Andrew Allibone, geographer Martin Doyle, and myself.

Vanda Station was one large green hut with a kitchen and small recreational lounge area. Around it were a handful of small portable huts that had one or two sleeping bunks in each. The loo, affectionately called "the honey pot," was a 44-gallon drum with a hole cut in the top located only a short walk away from the station, which in turn is located not far from the shores of the lake.

Lake Vanda is a stratified lake. The denser layers of salt and chemical-rich waters at the bottom have a higher temperature due to the magnifying glass effect of the four-meter thick ice sheet over the top of the lake. During times of high evaporation, when the lake dried up, the chemical-rich brine remained as a residue. As new melt water came in from the glaciers, the thick salty solution at the base did not mix with the incoming purer water, resulting in differential layers of differing density within the lake. Lake Vanda is approximately 70 meters deep and in places the water temperature may be as high as 113°F, although at the top of the lake it is just below freezing. The only life found in the lake are some microscopic algae that live between certain layers of water within variable depths of the lake.

Lake Vanda, Lake Vida, and Lake Vashka were named after huskies. Vanda Station was set up primarily to monitor the rate of water flow of the Onyx River, carry out meteorological observations in the Dry Valleys Region, and act as a general base for other studies of geology or biology in these regions.

At about 2:30 p.m. we were airlifted by chopper into the rocky flats surrounding the frozen Lake Victoria, close to the forbidding steep ice wall that marked the front of the Victoria Glacier. We put up our polar tent, tucked our gear inside, and then headed off for a long walk to Sponsor's Bluff where I could distantly see some craggy peaks of Devonian sedimentary rock. We climbed up to the saddle between two peaks, Sponsor's Bluff and Mt. Nickell, where we found an area of light-colored cross-bedded sandstone, rocks of the Beacon Group. It snowed continuously while we were there. From high up on this saddle we could look down into the pristine rocky wilderness of the Barwick Valley, where black metamorphic basement rocks were punctuated with jagged snow-capped peaks of cream-colored sandstones. It reminded me of my good friend and colleague Dr. Dick Barwick, now at the Australian National University in Canberra, after whom the valley was named.

Dick was one of the early NZARP veterans who worked as a biologist on the 1957-60 expeditions, so he was one of the first people to explore the region. He now works on fossil fishes with Professor Ken Campbell. During my two years in 1984-85 as a postdoctoral fellow in Canberra, Dick used to tell me hairy stories about those early days working in Antarctica. The helicopter rides were so dangerous in those days, he once told me, that the men would sit near the open doors with their full survival gear on ready to jump out in case of a crash.

We examined the rocks closely, getting nose-down onto the outcrops for any little fragments of bone that might indicate we had a potential fossil site. Although we searched intensely for some time, there was no sign of fish or plant remains, only some trace fossils such as the burrows and feeding trails of some ancient community of invertebrates, possibly shrimp or crab-like creatures. We made it back to camp around 11:00 P.M., in hazy light brought about by the closing-in cloud cover. We had a quick feed then went to sleep.

I lay there for the first time, snugly inside my tent on the Antarctic mainland, but all around me outside were rocks and gravelly sand. No snow or ice apart from the frozen surface of the Lake. Chris and I shared a polar tent in traditional layout, the two mattresses in parallel with a central cooking area between them. Andrew and Martin decided to sleep outside in low hollows on the ground, surrounded by a makeshift rock shelter against the prevailing winds, as dry valley areas like this are the only place on the mainland of Antarctica where one could actually get away with sleeping outside. They both wanted to experience a night in the open, under the cloudy sunshine.

On the following day we had mostly bad weather due to increasingly strong winds, so we were holed up in the tent for much of the day, only venturing out for short walks every so often. We spent quite a lot of time reading books and making cups of tea. Everyone was anxious to get on with our work and start the long trek back to Vanda Station, about thirty kilometers away. The plan was for Andrew and Martin to stay on a few days longer in the area to take measurements on the glacier and examine basement rocks in the region, then they would walk back to Vanda, studying the geology all the way back. Chris and I would depart with our backpacks loaded with food and water, and walk down through the valleys to Lake Vanda. Our most worrying concern was about the weather holding out for the long trek. We had a small portable tent on hand to escape the winds if it suddenly turned foul on us. The large polar tents and camping gear would remain at the Lake Victoria campsite and be lifted out by helicopter later that season.

On 29 December Chris and I set out on foot for Vanda Station at around 7:30 A.M. Our walk took us through the scenic Victoria Valley to the McKelvey Valley, through Bull Pass and into the Wright Valley where Vanda awaited us. We reached Vanda at around 6:05 P.M. with very sore feet from walking in Sorrell boots all day.

It was an amazing day's walk through some of the most alien landscapes I'd ever seen. Weirdly shaped granitic rocks (called "ventifacts") sculpted by the abrasion of strong winds littered these valleys in all shapes and sizes. The scene appeared even more surreal because of the intermixtures of different-colored geological materials, such as black doleritic rocks, light sandstones, and shining granitic pebbles, some lightly moistened by melted snow. Here in the Dry Valleys one sees sights that regular folk would never imagine existed on the continent of Antarctica, such as drifts of wind-blown sand built into monstrous dunes nearly 30 meters high, reminiscent of classic pictures of the Sahara Desert (at Bull Pass), or of clear flowing streams of ice-cold glacial melt water forming small rivers.

The Onyx River in the lower Wright Valley is Antarctica's only river that flows for about 60 days each year during the peak of summer. One of the other common sights of the region is the half-skeletonized mummies of desiccated seals buried in sand. One theory is that the seals get some sort of ear infection, lose all sense of direction and go wandering up into these valleys until they die of starvation. As there are no scavengers to prey on the carcass it quickly freezes solid after death. Then the harsh winter winds and blasting sands abrade one side of the carcass until the bones are cleaned of flesh. You can see the seal's skeleton on the side of the prevailing winds, and on the other side the animal is filled out with grey-black frozen flesh, its glassy eye and whiskers still intact. It's an eerie sight and one that reminded me of the continuous presence of danger from rapid weather changes. The weather for our walk was sunny and clear when we started off but it snowed lightly for most of the morning, and then cleared again after lunchtime. We arrived to a hot cooked dinner at Vanda Station, a couple of beers and beds to sleep in.

The next morning I arose around 11:00 A.M., as I was very tired from the previous day's long walk. After lunch I went for a short walk to the Upper Wright Valley, heading right of the huge table-shaped structure called the "Dais." I left Vanda at about 1:00 P.M. and returned at about 6:30 P.M. The day was spent pleasantly wandering around alone in the huge rocky valleys, exploring the geology and just getting my bearings in the overall area. I came across a lake of poisonous water that exists down in the valley on one side of the Dais. Apparently it was made toxic by the long-term concentration of residual minerals from the granitic and volcanic rocks of the region.

That night Errol, one of the meteorologists who regularly recorded the weather of the Dry Valleys, cooked us a culinary masterpiece for dinner. Where else but Cafe Vanda could you savor a meal of piping hot red saveloys (an Australian form of the hot dog, without the bun) in a tangy sweet and sour sauce made ingeniously from packets of powdered lemon and orange drink? It never ceased to amaze me what some people could create from dehydrated food supplies down in Antarctica. Martin and Andrew arrived from their long march home just after we'd finished supper. Now there were seven of us here: Dennis Corbett (Vanda Station leader), Martin Doyle, Andrew Allibone, Chris Rudge, Errol, and Ian (meteorologists), and myself.

On New Year's Eve Andrew and I set out on a long walk up to the Odin Valley. This was a high plateau where Devonian age rocks sat on top of the older Ordovician igneous rocks. Like many of the geographical features in the Dry Valleys, Mount Odin and its valley were named after Norse mythological characters. It gave the region a kind of Lord of the Rings mystical feeling about it.

We set off directly up the rubbly slopes of the Asgaard Ranges at 11:30 A.M., and it took us only three hours to ascend the 1500 meters up the rubbly scree slopes until we were standing on the rim of the Odin Valley, looking down over the Wright Valley. From up here Vanda Station appeared as a small green speck on the rim of the glimmering frozen lake. Getting down to work, I spent the next few hours carefully examining every outcrop I could find of the buff-colored sandstones (appropriately named the Windy Gully Sandstone) and searched carefully through the overlying shales, but found no signs of fossil bone. Above this shaley layer was another sandy unit of rock called the New Mountain Sandstone. Its spaghetti-like appearance was due to its incredible density of fossilized burrows, known as *Heimdallia*, possibly made by some ancient crustacean that lived nearly 400 million years ago.

We trudged up from the Odin Valley into the valley of the Heimdall Glacier, and then slowly made our way back down the steep soft slopes, treading carefully as we would occasionally lose our footing and slide down on the scree. The scree consisted of loose blocks of rock of all sizes, held together by frozen soil of smaller particles. Occasionally the scree would give way under foot because of these icy slip-planes. It's not hard to imagine twisting an ankle on such steep slopes if you were not careful.

At one time that day I heard the soft vibrating noise of a far distant helicopter as it came closer and closer, but despite scanning the skies all around me I couldn't see it anywhere. Suddenly, without warning, the big orange chopper arose immediately in front of me from behind the lip of the valley below and soared noisily over my head. I hadn't thought to check down in the valley where the helicopter was cruising below me all the time. I waved at the pilot and he waved back before quickly disappearing behind me over the top of Mt. Odin. It was an exhilarating moment as the chopper seemed to come out of nowhere with its low engine roar reverberating across the valleys, echoing back and forth.

That night we settled in for some traditional New Year's Eve drinks at Vanda Station, imbibing the "drambers" (Drambuie liqueur), a favorite drop of the Vanda meteorologists. For many years there had been a friendly rivalry between the "Vandals" (meteorologists) and the "Asgaard Rangers" (hydrologists and surveyors). The Asgaard Rangers would roam hill and dale carrying out their observations and measurements while the Vandals worked primarily around the station, doing meteorological recordings, letting off brightly colored weather balloons, and so on. Occasionally the bold Rangers would sneak down in the middle of the night to play some prank on the Vandals, like steal their flag or set up a booby trap outside the door. The Vandals would then retaliate by spotting the Rangers camp and carrying out a dawn raid on them. The stories are legendary, but when we were all sitting around together, in the warmth and solitude of Vanda Station, sipping the official drop of both Vandals and Rangers, all rivalries were momentarily cast aside.

At just after midnight Dennis, Chris, and I became the first members of the Lake Vanda Swim Club for 1989. We stripped off naked and ran quickly into the near frozen melt waters around the edge of the lake. It was bloody cold, but we were in and out very quickly and then hurriedly dressed again, so the effect of the water wasn't too distressing. The thought then struck me that no one will ever believe I did this, so I went back inside and grabbed my camera, then stripped off again and plunged in for a second time. I managed to get Dennis to take a somewhat fuzzy snapshot of me half submerged in the lake, modestly hiding my private parts just below the water line. We all went back to warm up inside the station. I was duly presented with Lake Vanda Swim Club membership certificate number 57, and then we all turned in to bed at around 2:00 A.M.

New Year's Day was a relatively quiet day. I decided to go for a walk up the valley and study the rocks of the valley floor, searching for traces of the fossil-bearing Aztec Siltstone in the glacial moraine. At the head of the Wright Valley the pyramidal shape of Mt. Fleming loomed above the Dry Valleys. As Mt. Fleming has a thick succession of Aztec Siltstone, I had hoped that maybe some of the fossil-bearing rocks had been transported into the valley during times of high glacial activity. Despite looking intensely all day I found no traces of such rock.

On 2 January I went for another long walk up to the Dais, the gigantic flat table-like mountain that sits in the middle of the Wright Valley just up from Lake Vanda. I climbed over the top of the Dais and down into the other side of the valley. This gave me an excellent view of Mt. Fleming, one of the best fossil fish sites in the region. I gazed at it longingly, and could almost smell the sweet perfume of long dead fish wafting off its icy peaks (sigh). In the afternoon I went to see Don Juan Pond, one of the other toxic lakes on the side of the Dais, then back to Vanda Station, after ten hours of solid trekking.

The plan for the following day was to pick us up and take us back to Scott Base. That morning we prepared all our gear ready for the helo pick-up. While waiting I went for a short walk around the base and looked at the equipment set up by the hydrologists for measuring water flow into Lake Vanda. The chopper arrived on time carrying American VIPs from the National Science Foundation, but as the weather had suddenly turned bad plans for our return to base were cancelled. The helicopter was securely anchored to the ground, and the visitors and chopper crew were forced to stay overnight at Vanda.

The next day the VIPs left with Chris Rudge, but there was not enough room on the helicopter for Andrew and myself, so we spent the day waiting around for another helo lift. After hearing later that no more choppers were coming, we went for another short walk in the Wright Valley. That night we heard on the radio from Scott Base that Margaret's team had been successfully pulled out of the Darwin Glacier region and were now back at the base. This was exciting news for me, as it meant that there was still hope I could get out to the fossil fish sites in the Transantarctic Mountains of the Skelton Névé region, if only I could get back soon to Scott Base!

On 5 January the helicopter came in and flew us back to Scott Base, using a rather indirect route in order to complete a couple of other missions in the region. Consequently we spent about five hours flying around or waiting on the ground for various pick-ups. We reached Scott Base at 4:30 P.M., exhilarated by a full day of chopper flying in and around all the dry valley regions. That evening I met up with Margaret Bradshaw and her group and heard all about their adventures on the two-month field trip in the southern Cook Mountains and Darwin Mountains. They had had a most successful trip in locating some new exposure of the Aztec Siltstone in the Cook Mountains, with fish fossils present. They also discovered more of the Derycke Peak meteorite on the top of that mountain, and had to modify one of their boxes and attach skis to its base so as to drag the heavy iron space rock down the slope.

From 6 January through to the 10th, my days were spent waiting around Scott Base, planning initially to go out in the field as soon as Margaret's group were ready for another field session. Eventually, as the bad weather settled in and all operations were starting to wind down, it soon became apparent that VXE-6 Squadron were not keen on putting us into a remote location at so late a time in the field season. We learned that VXE-6 Squadron had successfully recovered an old C-130 Hercules that season which had crashed near Durmont D'Urville in 1971. The plane (call sign XD-03) was retrieved and repaired and was used in active service for another ten years, right to the end of VXE-6 Squadron's replacement by the Air National Guard in 1998.

On Sunday, 8 January Margaret, Ray, and Eric Saxby went out cross-country skiing and convinced me to tag along with them. They taught me how to cross-country ski. We went about six kilometers out towards the ski field, along fairly easy level ground which skirted along the edge of the island. Although exhausted on return I felt proud at having covered so much ground on my first skiing attempt. The next day we went for a much longer cross-country ski to the Scott Base "Chalet"—about a twelve-kilometer round trip. It was decided that day if we couldn't get lifted into the field by Thursday we would have to call off the whole trip and start planning for the next season. Eventually the sad news of a helicopter crash suspended all operations and we had to completely abandon the idea of getting into the Boomerang Range. In such cases where lives are seriously at stake, one doesn't dispute the wisdom of the operations managers.

On the night of Wednesday, 11 January I departed Antarctica from Willy Field on a C-130 Hercules flight back to New Zealand, arriving in Christchurch early Thursday morning. By that evening I was back on a flight to Australia, somewhat disappointed at not having had the chance to visit the richly fossiliferous sites in the Aztec Siltstone, yet still exhilarated at my overall experiences of having explored the Dry Valleys on foot. Moreover, I was determined to go back and do it properly next time, as Margaret had promised we would plan a big return trip for a few years down the track.

And this is exactly what happened in the 1991-92 field season when we visited the "Mountains of Madness."

8

Back to the Great White South

At last everything was in readiness. The hour had arrived towards which the persevering labor of years had been incessantly bent, and with it the feeling that, everything being provided and completed, responsibility might be thrown aside and the weary brain at last find rest.
—Fridtjof Nansen

These immortal words of Norwegian polar explorer Fridtjof Nansen are an epitaph to the success of every great Antarctic journey, as he testifies that it is the years of hard work preparing for a polar journey that are the most difficult part. Once you have gathered your equipment, completed your training and are down there facing the forces of nature, you are at the mercy of your own past actions, as only the high degree of readiness achieved through your meticulous preparations can save you. Without adequate preparation and training, an expedition could easily turn into disaster, as so much depends on having reliable, safe equipment and a party of people who are experienced, well trained and level-headed during times of stress.

A lot had happened in my life since that first trip to Antarctica. In late 1989 I had been appointed as the new curator of vertebrate paleontology at the Western Australian Museum, so we moved from Hobart to Perth. The busy task of settling my family into a new life in Perth and familiarizing myself with the new duties of working in the museum had taken my mind off the possibility of returning down south to Antarctica. The scheduled return trip was planned for the 1990-91 season if the full logistic support we required was given. Unfortunately that was a particularly busy year for NZARP deep field trips, so our event (still called K221) was bumped off to the following year. Then one day I had a letter out of the blue from Margaret saying that our Antarctic event had received all the logistic support we asked for and was now on line to take place in late 1991. So, I had to get myself ready and be there later that year.

I decided that rather than depend on the National Geographic Society's generosity to get me back down to Antarctica the second time I would try Australia's Antarctic Division for support, seeing as my field area was actually located within the Australian Antarctic Territory. I managed to score a modest grant that enabled me to cover my costs of transport to and from New Zealand, pay for base and food costs, and I could further save on clothing hire by getting my kit through the Australian Antarctic Division. Furthermore, this meant I now had an ANARE event number (A136) that elevated our expedition to the status of being an international cooperative event between Australia and New Zealand. I must admit it felt good to go back to Scott and McMurdo bases proudly wearing the yellows of an Australian ANARE expeditioner, who were few and far between on those bases. I really did stand out in the crowd there between the hoards of red-clad USAPers and blue-clad NZARPies. However, never one to chance discomfort in the field, I was eventually talked into hiring some additional items of New Zealand gear to enhance my ANARE kit.

The new gear comprised a heavy survival jacket, a pair of leather double climbing boots, and a pair of NZ fleecy overalls (salopettes). As I had to go down to Hobart for a Gondwana 8 Conference held at the University of Hobart in June 1991, this gave me the opportunity to slip away one lunchtime and get kitted out at the Antarctic Division. I was amazed at how the Australian kit differed from the New Zealand gear that I'd used on my last expedition. The heavy outer clothing, boots, socks, and long johns were all very similar, only different in color and brand name, yet the amazing thing was the inclusion of the most unusual fashion accessories: two Antarctic Division ties, which one was meant to wear to any "formal occasions that might arise"; a crimson ANARE issue towel; and a standard issue sewing kit. The last item was perhaps the most useful thing I took to the field and was used every evening later on in the field trip to sew up tattered gloves and socks and do minor repairs on torn clothing. Even now, almost ten years later, I still use that little ANARE sewing kit for all my tailoring needs.

As I had already been to Antarctica once before and had completed the weeklong training course at Tekapo, I was excused this time round from participating in the course. Instead, on arrival in Antarctica our group would undergo a special intensive "deep field training course" down on the ice.

In late October 1991, once more I flew across to New Zealand and arrived at Christchurch, ready for the ultimate field trip, a three-month expedition to the remote wilds of the Transantarctic Mountains. We had the regular delays when flights were cancelled due to bad weather or delayed due to other reasons, so we passed time researching material on Antarctic fossils and geology and just hanging about in Christchurch.

Incidentally it was during my time waiting in Christchurch, on 25 October 1991, that an historic flight took place down south. For the first time, on that day, an all female crew of VXE-6 Squadron took a Hercules from McMurdo Base to open up the US Base at the South Pole.

We woke up at 3:45 A.M. on Tuesday, 29 October to get ready for our pick-up an hour later. This gave us time to have a filling breakfast of porridge and fruit, and then don our multiple layers of Antarctic clothing. We arrived at the Christchurch airport before sunrise and then waited outside with a long line of personnel for the boarding call. Just after the sun rose over New Zealand we climbed on board a dark green C-141B Starlifter, the scaled-down military version of a jumbo jet. By 8:00 A.M. we were cruising down the runway, pointed southwards, for our scheduled take-off.

The Starlifter contained 110 people on our flight. The back half of the plane was loaded with our baggage plus miscellaneous cargo for the bases. Although it was a military flight or "mission" as it's called by VXE-6 Squadron, in some ways it wasn't too unlike a typical commercial international flight, just like the Herc flight three years earlier. This time round, though, we were issued with boarding passes. A boxed lunch, almost identical to the one I'd been given three years before, was handed to us on boarding. The in-flight service wasn't too bad, although I was disappointed that there wasn't a screening of *Scott of the Antarctic* on the flight down. I won't say much about their frequent flyer points scheme, except to add that if ever I wish to go to the next Operation "Desert Storm," I'm almost there!

We huddled together on small canvas seats, facing the center of the plane in two rows. Bearing in mind that each expeditioner had to have a full set of survival gear close by them, mandatory in case of a plane crash, everyone lugged a large kit bag around their feet that left little room for moving about the cabin. As for toilets, well, we were told it's best to go before the plane took off. However, in an emergency there was a small funnel attached to a tube down the back of the plane, with a short draw curtain around it. Very blokey indeed. No little lights came on when the funnel was vacant either! I found this to be perfectly okay when in absolute dire desperation for a leak. Any other combination of gender and bodily relief did not accord with the C-141B funnel system, although I later found out that there were other ways to relieve oneself within the plane's cargo hold area. I asked for no more details and, mercifully, on a Starlifter it's only a 4.5-hour flight from New Zealand to Antarctica.

Later on during that flight we were each allowed a quick visit to the cockpit to admire the extraordinary view. We had just reached the edge of Antarctica and could see the northern part of the snow-covered Transantarctic Mountains, with the vast polar plateau stretching out as far as we could see. From the main cockpit the 180° view was simply magnificent.

Standing next to the female pilot in command of the flight, I looked out across the top of the world, which to our regular antipodean view was actually the bottom of the world! Quite a mind blower. I gazed in silence at the magnificent rocky snow-draped mountains below fringed by the frozen sea and pancake ice structures, on one side, and by an apparently infinite plateau of polar ice on their far side. Above us the boundless blue sky was punctuated with a few fluffy cirrus clouds and the sun shining down on all of it. The latter was probably the most significant factor in the whole scenario for us, because if the sun disappeared, the threat of losing visibility would force us to turn around and go all the way back to Christchurch.

The plane touched down about 2:00 P.M. on the ice runway of Willy Field. After gliding along the ice on its skids for quite a while, it finally turned around and taxied in to its allocated standing area. By then the internal temperature on the plane had been gradually turned down so as to acclimatize us to the cold environment into which we were about to step.

Some of the old, wizened expeditioners had told us horrific tales of the first few flights down in which the temperature was kept warm and cozy all the way to Antarctica. Then, on opening the doors and stepping out into –31°F and freezing winds, some of the passengers had the fillings in their teeth contract rapidly and pop out! So, fully clothed and expecting the onslaught of cold air, we descended from the front door of the plane with mouths clenched, to find a pleasant Antarctic afternoon with a temperature of about 10°F, and no breeze to speak of.

We were greeted on arrival by our field survival leader, Brian Staite. Brian is a swarthy bearded Kiwi, stocky and of robust build, with a quietly competent attitude. He smiled at us and came forward to firmly shake my hand, as I was the only member of the field party whom he hadn't already met. I smiled back and assessed him. We were soon going to share a polar tent together for two months. He seemed like a decent sort of fellow and was a veteran of several long deep-field Antarctic expeditions. His background in mountain climbing, skiing, mountain rescue and other outdoor pursuits qualified him as one of the most skilled of all the New Zealand field leaders. My estimation of him only grew over the next two months as I got to know him better. I could not have asked for a better mate to share a tent with on a journey through Antarctica.

We were driven to Scott Base and given a quick briefing with the base manager, Phil Robbins. He warmly welcomed us, gave us the lowdown on base rules, then showed us to our temporary sleeping quarters. At the time there were many expeditioners staying on the base, so we were relegated to the old quarters. These were the original Scott Base living quarters used in the early days of the NZARP before the flash new base was erected. They consisted of a large open interior space with various double bunks, so we all shared the room with about five others. Privacy is a rare luxury in Antarctica on base mainly because there just isn't the room unless you are one of the winter-over crew. You are then allocated a small room each, as having your own space is very important for your long-term sanity over the cold, dark winter months.

Most expeditioners spend as little time hanging around the base as possible, as their main aim is to get out and work in the field as soon as logistics permit. If your logistic support was a helo flight this may not take long at all, maybe only a few days, depending primarily on the weather. However, we were special because we were one of only two deep field expeditions that year, so we had to stay on base longer to prepare ourselves with extra training, including getting familiar with driving and adjusting our skidoos, as well as testing out all the routine equipment like the radios and stoves. The best way to do this was practical experience. We planned to do a couple of shakedown trips to make sure that all of this vital equipment was in perfect working order, and at the same time to see some of the local historical sites on Ross Island.

I will never forget the day I joined the Polar Plungers' Club, on 7 November 1991. I had no desire to go swimming naked below the frozen sea ice, but I really had no choice. A day after I arrived on base I found out that Australia had beaten New Zealand in some major rugby competition and, being the only Australian on base at the time, friendly rivalry was mounting against my kind. I was issued with the challenge "Come on Aussie, join the Polar Plungers Club or else sign the wimps' book" ("wimps" being pronounced emphatically without any vowel). So, what else could I do? I had joined the Lake Vanda Swim Club on New Year's Day 1989, so I thought that this would be a fairly similar experience.

The next afternoon those of us who were about to be initiated into the Polar Plungers' Club, four in all, walked on the sea ice outside the base to a little hole that had been cut into the ice for the marine biologist divers. The water was continually freezing up, so each day teams of workers kept the hole open by shoveling out the partly frozen sludgy ice layer on top. Next to the hole in the ice was a small mobile hut, mercifully fitted with a heater, where the divers would change after each stint underwater. We all entered the hut and stripped off, wearing only our boots and heavy survival jacket. Next we all marched out in single file to stand near the open hole. By this time hordes of spectators had arrived and were perched around the site with videos and cameras, like voyeurs at Roman amphitheatres eagerly waiting to see the Christians being fed to the lions.

I was first up as I had the dubious honor of being an Australian. Proudly, I took off my jacket and stood there naked in my boots, showing them all what we little Aussie bleeders are really made of! The outside air temperature was –13°F and a light breeze was blowing. Never before had I experienced such a helpless feeling of the human body being so frail, so vulnerable to the forces of nature. They fitted me with a harness around my chest, and then pointed to the hole with its slushy, dark blue frigid water. Looking back at the video recording I saw that I hesitated slightly, also pointing at the water as if to say "You mean you want me to jump in there?" Then I leaped in.

All I can recall of that brief second or two I spent underwater is opening my eyes and seeing the most amazing scene of blurry, blue icicles hanging down from under the sea ice, and some menacing dark shapes moving around. I was pulled out of the water then onto the ice where I immediately started jumping up and down on top of my boots. A towel was handed to me, I dried my feet first, hurriedly put on my boots, and as quickly as possible hopped over to the hut to dry off and get dressed. It felt great, though, an invigorating shock to the system.

As I wandered back to watch the others go in after me, I discovered that everyone mustered around the ice hole was laughing their heads off.

"What's the matter?" I asked, "it couldn't have been *that* funny!"

Fraka, who had been video recording the event, was practically in tears of laughter. She told me that only a second or two after they'd hauled me out a large Weddell seal had leaped up out of the same hole onto the sea ice, startling all the onlookers.

"Whatever was it after?" she jested. I just winced and walked away silently.

Perhaps Percy Correll, one of the men on Mawson's expedition, was the very first man to inaugurate the Polar Plungers' Club. When unloading the *Aurora* at Adelie Land, the men were shifting cases from the boats to the landing stage and a case of jam fell into the water. Mawson ventured in the water and took some time to extricate the box and keep his head above water, but a few days later he observed one of his men actually voluntarily taking a dip in the sea. Mawson recalls his experience:

> At last I went in, and standing on tiptoe, could just reach it and keep my head above water. It took some time to extricate it from the kelp, following which I established a new record for myself in dressing. . . I do not think I looked very exhilarated after this bath, but strange to say, a few days later Correll tried an early morning swim which was the last voluntary dip attempted by anyone.

9
Cape Evans and Cape Royds

Cape Evans is one of the many spurs of Erebus and the one that stands closest under the mountain, so that always towering above us we have the grand snowy peak with its smoking summit. North and south of us are deep bays, beyond which great glaciers come rippling over the lower slopes to thrust high blue-walled snouts into the sea . . .

Ponting . . . declares this is the most beautiful spot he has ever seen . . . .

—Robert Falcon Scott

When the mercury hovers between 21°F to 36°F in Antarctica it's positively tropical! This was the case on Saturday, 9 November, so we decided to take advantage of the great weather and do a shakedown trip on the skidoos to Cape Evans and Cape Royds to visit the historical huts and see the Adelie penguin colony. We straddled our skidoos about 9:30 A.M. and rode straight to Cape Royds. There was virtually no wind when we set off—ideal skidooing conditions. We scooted passed Cape Evans, deciding to get as far as we could while the weather was fine, and then slowly make our way back to Scott Base stopping at other sites of interest along the way. We arrived at Cape Royds about noon.

Cape Royds is the site where Shackleton's hut was erected for the 1907-09 British Antarctic Expedition. It is situated on a prominent black volcanic rock outcrop that is home to many thousands of little Adelie penguins. These penguins grow to about 70 centimeters high, and have black faces with white rings around the eyes, black backs, and white bellies. At the time of the year we visited them they were busy making nests for the mating season. We watched the males carrying around tiny rocks in their beaks and placing them delicately to make a ring of stones for the female to nest in.

Adelies are one of only two penguin species, the other being the emperor penguin, that breed on the shores of the Antarctic continent. Adelies spend their winters in the pack ice out on the sea where temperatures are slightly warmer than on the land. In spring the male Adelies make long journeys over the sea ice to return to their colonies, like the one at Cape Royds, where they diligently hunt for their old nests and refurbish them for the new mating season. The females then arrive and help their mates with the nest building. The eggs are then incubated for about a month and the young brooded for about four weeks. The interesting thing is that since their arrival back at the colony to renovate the nest and undertake courtship, the males have not eaten for almost three weeks, and after the female lays the eggs, the males take the first incubation watch. They lose nearly half their body weight before being able to have another feed. From here on the males and females alternate every three days or so, feeding at sea and bringing back food for their young chicks in their crops.

These beautiful little penguins seemed to have no fear or concern at all about us being there. By walking carefully amongst the penguins, then sitting down still, I could watch them very closely as they would happily carry on their work walking around me. It must have been easy for the early explorers to procure penguin meat whenever they needed to. In many cases it supplemented their regular diet or was a major part of it during times of extreme isolation away from base. Victor Campbell's Northern Party of Scott's 1910-13 Terra Nova expedition became stranded away from the base as winter closed in upon them. The six men lived for over six months in a fairly small ice cave, subsisting largely on penguin and seal meat, although not without some gastric problems. Here is a quote from Campbell's account of one mealtime:

> We made a terrible discovery in the hoosh tonight: a penguin's flipper. Abbott and I prepared the hoosh. I can remember using the flipper to clean the pot with, and in the dark Abbott cannot have seen it when he filled the pot. However, I assured every one that it was a fairly clean flipper, and certainly the hoosh was a good one.

We watched the Adelie penguins for about an hour and occasionally observed some interesting behavior. One lone little penguin, well away from the rest of the crowded rookery, came in tobogganing on his belly across the ice, then popped upright onto his feet when he reached the volcanic rock. It was amazing how fast he could scoot along like this. It does make you wonder why he went out by himself in the first place when all the action was happening at the rookery, but I suppose he must have had his own penguinine reasons that we mere humans could never comprehend.

Shackleton's hut stood proudly on the rocky bank not far from the penguin rookery. It commanded wide sweeping views of the oily blue Ross Sea with its huge floating icebergs. One very large iceberg was still there from Shackleton's day as captured in the photographs of the 1907-09 British Antarctic Expedition.

The hut is now beautifully restored with all the items inside in actual position, based on the original photographs. The old sledges hang from the roof, and the items of food are still out on the shelves made from packing crates. I examined a can of Aberdeen Marrow Fat and thought to myself that they really must have been desperate to eat such high cholesterol food to survive. Only later did someone explain to me that this was merely a brand name for green peas. In many cases the early expeditions required much funding to get started and any sponsorship for food was welcomed. On Scott's expeditions his men were often photographed either scoffing down a well-known can of baked beans, or holding up a certain product for future endorsement. The other items of food left in the hut included many packets of hard plasmon biscuits, the staple diet of sledging parties when mixed with pemmican to make up a "hoosh." One packet was open on the table with a few broken biscuits displayed. I presumed that this was for those who wished to try the food, so I broke off a very small fragment and ate it. It tasted like hard, flavorless cardboard. Nonetheless, on a long sledging journey, the biscuits mixed up in the hoosh would have provided valuable carbohydrates and dietary fiber.

Outside the hut the remains of the "car" can be seen along with the makeshift kennels where the dogs lived, now filled with ice and snow. Snow-covered bales of hay, used for feeding the ponies, are neatly piled up near the kennels. It must have been a harsh environment for the animals, but they were breeds adapted to the cold and rarely suffered during the early expeditions. Raold Amundsen took 97 dogs on his ship *Fram* when he set sail for Antarctica but he records that during the first week of their sledging journey to the South Pole, on Friday, 15 September 1911, it was so cold that two of the dogs froze to death after they lay down.

Unlike Scott's two huts, Shackleton's hut at Cape Royds is small yet gives the impression of having been better organized. Much has been written on the differences between these two great men and although I am no expert in matters of historical perspectives I did have a feeling that this hut had a happier, less military coziness about it than either of Scott's huts. All ranks were in together here. Shackleton didn't like to alienate himself from his men; he wanted to share their space. Scott, in all fairness, came from a naval background where it was more or less expected at the time to house the officers in separate quarters from the other men. After an hour or so of examining the hut and its contents we carefully closed up the shutters over the windows, locked the front door, and silently stepped outside into the 1990s once more.

Outside in the sea ice in front of the hut was an enormous berg, a gargantuan milky white crystalline mass of ice, framed in all its grandeur by a pale blue sky above and deep blue sea around its girth. The black volcanic prominence was alive with myriads of busy little penguins, looking somewhat akin to swarms of formally attired ants from a distance away.

Our next destination was Scott's hut at Cape Evans, the base from which he and four of his men left for the South Pole in November 1912, never to return. Cape Evans on a perfect day is a magical site, as vivid and powerful a place as Scott described it. The huge smoking cone of Mt. Erebus broods over the restored grey wooden hut. A big, black wooden cross rests on the low volcanic hill behind the hut as a permanent reminder of the bittersweet mix of bravery and tragedy that made that expedition so famous. In contrast, a short distance away from Scott's hut was a modern demountable base with a humming windmill harnessing the vast wind power of the region. This was the Greenpeace Base.

The first thing I recall about stepping into Scott's hut at Cape Evans was the huge scale of it. It was a big building with an outer covered stable area for the ponies. As I entered the building I recall bumping my head on the lintel, not because the door was so small, but because the build-up of snow had raised the ground level considerably higher compared to when Scott and his men were there.

As with Shackleton's hut, every artifact inside the Cape Evans hut was placed back in its original position, matching the interior scenes from Ponting's early photographs as closely as possible. The long dining table had a single large pewter mug at its head—Scott's mug. Behind there lay Ponting's darkroom where many of the earliest photographs of Antarctica were developed and printed. Wilson's biological laboratory was nearby with a stuffed emperor penguin placed on the table next to various dissecting instruments. In the corner around from there were Scott's modest sleeping quarters. His small, portable bunk still had his fur sleeping bags and various pairs of socks hanging up around it. It is quite a humbling experience to actually lie down on Scott's bed and stare up at the wooden ceiling in the same manner as he did some 80 years earlier, trying to imagine his thoughts as he arose on Wednesday, 1 November 1911, the day he set off for his march to the South Pole.

Once more, as I've said with respect to the other huts, the feeling of the moment, that these men were only here yesterday, is hard to suppress. If there are such things as "ghosts" in the scientific sense, then there must reside the essential memories of those men; their imprints in time hanging in the very ether of the hut's moody air.

We spent some time looking at the items, reading some of the old magazines they had left lying around, and just drinking in the dank atmosphere of the place. It was a fairly dark hut, deliberately having few windows so as to brace up to the strongest winds, but well enough constructed to survive the harsh winter onslaughts of some 80 or fiercer Antarctic winters.

All around the hut, I thought that actual parts of the men from Scott's expedition must now remain, somehow minutely integrated with the local environment. The sweat on the ground, the shed flakes of skin and frozen particles of their bodily excrescences, must still exist within or around that area. No dust mites or insect scavengers exist here to break down the microscopic organic debris shed by a large band of humans and their animals. I thought about this deeply one day and it dawned on me that here, more than anywhere on Earth, are to be found the forensic ghosts of past people.

In Noel Barber's book, *The White Desert*, he recounts an extraordinary meal he had prepared by Australian Antarctic veteran Sir Hubert Wilkins, who delved into a hidden cache of food that Scott had left in a hollow halfway up the hillside in case of an emergency, such as if the main hut burned down.

> I ate the strangest meal of my life, a meal cooked and tinned at least fifty years ago, but which due to the natural refrigeration of the Antarctic, was just as good as new . . . The meal was excellent and there were hundreds of tins of it. The cheese was rather high, and tended to crumble when we opened the tin, but it was quite edible. The biscuits still retained much of their original crispness. There were score of tins of English vegetables, some wonderful greengage jam—I tested that, too—boxes of Quaker Oats, Cerebos salt, and Coleman's mustard in huge tins and all in well nigh perfect condition. The only thing that had gone bad was some corned beef.

After leaving Scott's hut we walked up to Observation Hill to examine the cross erected in memory of all who died on Scott's last dash to the pole. The large cross was made of Australian jarrah* and had engraved upon it the immortal words of Tennyson's poem "Ulysses": "To strive, to seek, but not to yield."

We pondered this in silence for a few minutes, taking in the magnificent view of the whole area, then ambled slowly down to the Greenpeace Base.

A small Turkish man from Australia, nicknamed "Oz," was maintaining the base while the other members were out on various environmental observation missions. He warmly greeted us and invited us in for afternoon tea. This small but modern base was even more upmarket that Scott Base in its advanced technology. It had a kind of Swiss laboratory feel about it, brightly lit with sterile white fluorescent lights and displaying a preponderance of leafy green plants sprouting from the hydroponics corner. Oz treated us to fresh plunger coffee,

---

*An Australian gum tree noted for its durable reddish-brown wood.

chocolate biscuits, dried fruits, and nuts, and was more than happy to chat to us about the role that Greenpeace plays in Antarctica.

Right from their first arrival in Antarctica they had been monitoring the large American McMurdo Base to check that waste material was being carted away rather than dumped and that the local wildlife colonies were not in any danger from the expansion of human activities around the bases. Their job was simply to let the rest of the world know what went on down here environmentally. In addition to the fairly modern base they had set up at Cape Evans, they had a few demountable fiberglass "igloos" plus some polar tents and could camp out at various locations to carry out their observations. One such "advance base" was a little structure not far from Scott Base, from which even the New Zealanders were under their scrutiny.

In general terms relationships were good between Greenpeace and Scott Base, although the US military personnel and USAP scientists were officially told not to have any contact at all with the Greenpeace base. Nonetheless, we heard about several of the US scientists who had called in on Greenpeace for "unofficial" friendly visits during the time we were around Scott Base. It was one of the few places you could call in for a social visit, provided, of course, that you had your own transport.

Before departing from Cape Evans we collected a sample of the rock from the Mt. Erebus lava flow, a peculiar variety of basalt called "kenyite" (originally after a type of volcanic rock found in Kenya). On our way home we stopped at the famous ice wall of the Barnes Glacier where Scott's photographer, Herbert Ponting, shouted to hear his echo, and we all emulated him by yelling as loud as we could and listening to the echo reverberating off the distant hills.

A few kilometers further on we reached a long river of protruding ice known as the Erebus Ice Tongue. We found some large caverns and explored inside one. The light was a deep blue shining down through the thin ice ceiling above. This experience gave me two scary thoughts. First, it gave me the eerie feeling of what it must be like down the bottom of a crevasse. Second, by looking up at the thin ice cover above, I could see just how dangerous walking around on ice floes could be. Inside the ice cave I saw amazing interplays of subtle light on the jagged shapes of ice stalactites hanging down from the overhead slits. Frank Wild, the geologist on the main eastern journey of Mawson's 1912 expedition, expressed his feelings of joy and awe at seeing ice sculpture at its natural finest:

> At the actual point of contact was what might be referred to as gigantic Bergschrund: an enormous cavern over one thousand feet wide and from three hundred to four hundred feet deep, in the bottom of which crevasses appeared to go down forever. The sides were splintered and crumpled, glittering in the sunlight with a million sparkles of light. Towering above were titanic blocks of carven ice. The whole was the wildest, maddest, and yet the most grandest thing imaginable.

We jumped on our skidoos and raced back to Scott Base after this full day of exploring. It was exhilarating to ride the skidoos, two up, like sitting pillion on a motorcycle, almost flat out at about 16 miles/hour across the flat sea ice. These sleek machines are powered by 500 cc two-stroke motors so are capable of generating a considerable amount of power, especially as in this case where they were not pulling heavy sledges. Still, it is rare that you can drive the things with any degree of speed, as it takes almost perfect conditions. Then, the trouble with going fast is that you generate a significant wind-chill factor on your face and gloved hands, which soon becomes uncomfortable, and eventually forces you to slow down and put on extra clothing. The wind-chill is about 1 degree for each knot of wind speed, so even though it was quite warm at 23°F, at 16 miles/hour it drops your comfort level to about –22°F, which can be rather cold on the nose. We arrived back at base at around 7:30 P.M., relieved to discover that they had saved dinner for us, complete with some Australian table wine left over from some official function. It was a perfect end to what was arguably one of the most interesting days I had ever spent in Antarctica.

I kept myself busy the next day working in the Scott Base science lab on their Mac computer, writing a summary article on Antarctic dinosaurs for the Scott Base Times newsletter. That evening I tried to make a phone call back home to my family but gave the operator the wrong number. Instead I reached the Royal Melbourne Golf Club. When I told the man I was calling from Antarctica he sounded a bit doubting over the authenticity of the call and hung up. "Just another prank call from Antarctica," I guess he thought.

I tried again and eventually reached my family. It was wonderful to hear my wife and children's voices over the phone. I relayed the news that all was going well so far and that we were to be put into our field site within the next couple of days. After that I could not expect to hear any news from home for about a month into our trip, when we expected a helicopter to reach us with fresh supplies and any letters or parcels from home.

We were going on our recon flight the following day. I was excited by the prospect of getting a chance to see the lie of the land where we would be venturing. Most importantly, as I had good eyesight, I knew I could probably spot the characteristic mottled layers of the fossil-rich Aztec Siltstone. It would be a most interesting day.

10
A Flight of Discovery

However such preoccupations did not bother us then. We were on a flight of discovery, and wanted to see things and record them.
—Robert Byrd

Monday, 11 November was the day we were scheduled to do our recon flight over the field area. This was very important for the success of our expedition as it gave us the opportunity to closely examine our sledging route up the McCleary Glacier to see if there were any visible signs of crevasse fields, and to look at the mountains to get some appreciable idea of where the most accessible fossil-bearing rock layers could be found. The prime purpose of the flight, though, was for the benefit of the aircrew, as the VXE-6 pilots would not usually land a Herc in a remote location unless the landing spot had been carefully examined and photographed. The pilots were all very experienced at polar flight and, thankfully for everyone concerned, would take no unnecessary risks.

Our flight took off about 4:30 P.M. in one of the US National Guard Hercules. In 1998 the National Guard took over the job of VXE-6 Squadron after its 43 years of service to the US Antarctic operations. This trip in 1991 was during the time when the National Guard pilots were just getting their flying experience in Antarctica. It was a fine clear evening, with excellent visibility over the Transantarctic Mountains. We commenced our journey by flying up the Wright Valley, past Lake Vanda and over the twisted maze of rocks called the Labyrinth. We zoomed past Mt. Fleming, down alongside the Royal Society Range and then cruised slowly over the vast white snow plain of the Skelton Névé.

The vista of the mountains was stunning. They are made up of layered light-colored rocks of the Beacon Supergroup, composed of mostly whitish-yellow sandstones and greenish red shale, interspersed with deep black dolerite sills and wedges extruded randomly through them. In some places it appeared as if the extrusion of doleritic volcanic rock had rapidly eaten up the placid sandstones like an incurable cancer. These mountains were now the ancient sentinels guarding the secrets of Antarctica's rich prehistoric parade of past life.

As we flew over the Mulock Glacier with its enormous, regularly spaced crevasses, we headed south alongside the Cook Mountains towards the Darwin Glacier, at 80° south. Our chosen landing site for the expedition was to be in the middle of this glacier near some isolated rocks dubbed Roadend Nunatak.

We flew several circuits all around the glacier and over the fringe of the Darwin Mountains, carefully scrutinizing the landing site and searching for alternative sites in case of any last minute changes in plan. We were all flat out during the three hours of that flight taking photographs almost non-stop, checking the maps for accuracy as to topographic references and making detailed geological notes. Fraka also made a video of most of the flight so we could ponder over our sledging routes later in the comfort of Scott Base. There were many large crevasse fields down there, most of which were clearly marked on the topographic maps. As Brian was primarily responsible for our safety on the ground, I was pleased to hear that he had visualized a safe route up the McCleary Glacier, although he mused with a wry grin that the head of the glacier might be "a little tricky" due to a steep incline with ice falls on both sides.

On the flight back to base we flew over the Skelton Névé, identifying the camel-humped shape of Mt. Metschel, the towering Portal, mighty Mt. Crean, and the pyramid-shaped Mt. Fleming. From high up there it seemed like fairly easy going on the flat snowy expanse of the Skelton Névé. Crevasse fields were clearly visible close to any mountainous outcrops, as this is the normal result of massive volumes of hard ice flowing around an immovable object. Crevasses are merely pressure cracks developed in regular patterns perpendicular to the direction of the ice flow.

We flew into the Wright Valley and suddenly the Herc dropped a few hundred meters as it passed directly over the Airdevron 6 ice falls. The plane then cruised over the sea ice and came in for a soft landing at McMurdo at about 7:30 P.M. Back at Scott Base we quickly ate dinner then had a few beers in the bar to mull over what we had seen on the flight. The route up the McCleary Glacier was our main concern because at its origin, near Festive Plateau, it had only narrow access surrounded by cascading ice falls on either side. This meant that crevasses were certain to exist in the area, hidden from our aerial view by the recent snowfalls.

The next few days were spent hanging around the base waiting for our final put-in flight. Conditions once more had to be perfect for the flight. On Tuesday, 12 November we did a final check on our two skidoos at the garage. They were fairly new Alpine skidoos. The only problem we would have to be aware of was that at high altitudes the jets on the carburetors would need changing, so we were shown how to do the job. Then we had to do it ourselves to the satisfaction of the head mechanic.

On Wednesday, 13 November, I kept working on the computer in the science lab on an article for our museum dinosaur club magazine about the various dinosaurs that had so far been discovered in Antarctica. Back then there were only a few to speak of, and none had been formally named or described in detail. Ankylosaur bones from James Ross Island and some ornithopod bones from Vega Island, both on the Antarctic Peninsula, were the only published record of Antarctic dinosaurs, but we had heard on the grapevine about the discovery of a nearly complete meat-eating dinosaur skeleton from Mt. Kirkpatrick the year before. This was the *Cryolophosaurus*, which I mentioned at the beginning of this book, but at the time we knew virtually nothing about the beast.

That afternoon we made our expedition sledging flags. Every major deep field expedition does this. We planned our design and chose our colors. Two flags were made so that at the end of the trip the New Zealand team leader could have one and I would keep the other one for my museum. Each flag bore a design showing the outline of the Antarctic continent with a fossil placoderm fish (a *Bothriolepis*) sewn over it. The words "K221-A136 Devonian Beacon Studies" were written over the motif. They were in green and red with a yellow map of Antarctica. At every camp we would fly the flags from a tall bamboo pole to serve as a visual beacon of our site, and to determine prevailing wind directions. My flag was donated on my return to the Western Australian Museum history collections because it signifies the first official Western Australian Museum expedition to Antarctica.

On Thursday, 14 November, we received the good news that our put-in flight would be the first event scheduled by VXE-6 for the next morning. We spent the day finishing odd jobs, like our washing and writing our last postcards to home, and just generally relaxing, saving our energy for the big day ahead.

That evening we went for a short ride on the skidoos. It was –4°F outside and the wind-chill gave the air a bit of a bite. We all went to bed early that evening.

That night I rested uneasily. The excitement was too much to contain. We would be in the remote deep field for the next two months, at the mercy of the cruel Antarctic weather, and in the hands of fate as to whether we would actually find anything interesting, anything that might be new to science. I kept thinking how I was now locked into this extreme course of action, with no recourse to back out. The next day the journey would begin in earnest.

11
On Mr. Darwin's Glacier

We dwelt on the fringe of an unspanned continent, where the chill breath of a vast, polar wilderness, quickening to the rushing might of eternal blizzards, surged to the northern seas. We had discovered an accursed country. We had found the Home of the Blizzard.
—Douglas Mawson

The next morning we rose early, enjoyed a hearty breakfast, and rushed out to the airfield so that we could be there at 7:00 A.M. sharp as we were expecting the flight to get under way at about 8:00 A.M. However, due to some last minute hold-ups caused by minor repairs we had to do to the hydraulic system on our aircraft, we spent the next few hours just waiting around in a small, featureless room. At about noon one of the Scott Base staff came over with some lunch for us; then after eating it was back to waiting again. Finally, at about 1:30 P.M. we were told that the plane was repaired and ready for our mission, which was to be led by Flight Lieutenant Scott Allen.

The next job was to load up the Herc with all our gear. We had decided to tie it all down firmly on the sledges and load the skidoos and sledges straight into the belly of the plane, rather than follow the conventional way of loading everything separately then having to pack up the sledges when we arrived on the Darwin Glacier. Driving the skidoos up the narrow plank into the backside of the plane was a little tricky, but we eventually managed to get all of our gear on the plane without any hassles.

The flight over the Transantarctic Mountains took about an hour. Large crevasse fields were clearly marked by parallel rows of icy lines across the really wide glaciers. Mountains poked upwards out of the snow-covered plateau, gently draped in snow. Sunlight glistened off the exposed expanses of blue ice on the glaciers. Although the roar of the Herc engines precluded much talking in the hold, in the cockpit Margaret and Brian had radio headsets on so they could discuss the landing spots with the pilots. As we circled around the area we could see that the route we had chosen up the McCleary Glacier was going to be a challenge for there were several large crevasse fields along the way. We would have to stick closely to our planned route, which looked safe enough visually, but one could really never be sure until you were there on the ground. Crevasses have a bad habit of being lightly snowed over. This occurs when winds blow fresh snow across the crevasse and it freezes to form a bridge of thin ice. Later, after light snow falls blanket the whole region, the crevasse becomes virtually invisible.

At around 3:00 P.M. the big plane buzzed low over the landing spot and bounced upon the snow-covered surface of the Darwin Glacier field with its huge skids a couple of times to check that the area was free of hidden crevasses. It then turned around and came in to land. Landing on the ice is always a little dangerous as no one can exactly predict or know what the conditions will be like. If a large crevasse had been covered over by snow, the plane could suddenly crash down into it and the plane would cartwheel over on the hard ice. This has actually happened before in northern Victoria Land, so it is not an implausible scenario. Naturally we were all a little apprehensive as we touched the ice and slid along the wild runway. As the plane pulled up we sprang into action, untying the cargo and getting ready to unload all the gear out of the backside of the Herc.

As the tailgate opened we were instantly assaulted by both the noise of the roaring engines and the blast of freezing air they created. Within about fifteen minutes we had unloaded the sledges with all our gear. These just slid right out down the ramp onto the ice along with the skidoos. Our next job was to erect the radio aerial and test the radio communications. Once this was done, the plane revved up its engines and headed off. At about 4:00 P.M. we watched the black speck float into the distance, high above the lofty mountain peaks, heading towards McMurdo. I knew then that we would not have recourse to anyone other than ourselves for some time ahead. We had to get used to the idea that no matter what happened in the course of the next month or two, our radio was probably going to be our only contact with the outside world.

Finally, I thought, we're here. Inland Antarctica, at 80° south, on the open icy flats of the Darwin Glacier, named in honor of one of my great heroes, Mr. Evolution himself, Charles Darwin. I had wondered why someone would name a glacier down here after Darwin, whose major work on evolution and natural selection was founded primarily on the observations he made in tropical rainforests and isolated Pacific islands. Then I discovered through my reading that on Scott's first expedition they had taken only one book on their sledge journey towards the South Pole and one which they read every night: Darwin's *On the Origin of Species*. Hence the connection.

It was a sunny, clear afternoon, and the sky was a rich blue hung with a few puffy white cumulus clouds. We spent the rest of the day setting up a camp, getting acclimatized, playing around with our gear and just getting comfortable with our set-up. Stunning views of the mountain ranges surrounded us, some appearing as almost transcendental peaks extending into the far distance, others like shrouded biblical towers in a surrealist painting.

The peaks to the furthest south that we could see were the Churchill Mountains, almost two hundred kilometers away from our camp. We could only see the snowy tops of these mountains. They were so far away from us that for the first time in my life I was able to see and comprehend the natural curvature of the Earth.

It made me feel very humble to be able to do this.

12
The First Worst Day of My Life

At last we were truly entering the white, eon-dead world of the ultimate south. Even as we realized it, we saw the peak of Mt. Nansen in the eastern distance, towering up to its height of almost fifteen thousand feet.
—H.P. Lovecraft

The highest peaks to the south of us, such as Mt. Kirkpatrick near the Beardmore Glacier, do indeed rise to over 4000 meters, as Lovecraft has written. We shifted camp the next day from the Darwin Glacier to the base of a large mountain range that ran for nearly 30 kilometers, and was crowned by the towering peaks of Mt. Longhurst (2846 meters) and Mt. Hughes (2250 meters). One of the unnamed peaks closest to us was nicknamed "Gorgon's Head" by Margaret's field party in 1989 because it had black tendrils of doleritic volcanic rock twisted through the top of its lighter buff-colored sandstones and greenish-grey shale. It looked like the head of the mythical gorgon Medusa, and had about as much charm on a blustery day. The site produced some fragmentary fish fossils, so we had planned to make it our first destination.

I wrote down in my field diary that evening that this had been "the worst day of my life." I have since had others that would rate as far worse than this, but that will become evident later in the story!

It was a chilly 1°F when we woke up. We finished packing up the camp by 10:30 A.M. and were ready to roll. It was our first attempt in the field at lashing down the sledges tightly and preparing the skidoos for a full day's sledging. We were to head across the Darwin Glacier to the Cook Mountains, which lay to the north of our camp. Yet this was not a simple case of heading in a straight line for the peaks where we wanted to go. Large crevasse fields in the middle of the Darwin Glacier forced us to follow a less direct route, parallel with the glacier, and then we had to cut across diagonally towards Gorgon's Head.

Despite its off-putting name, Gorgon's Head is actually a treasure trove of fossils irresistible to the likes of us paleontologists. The first record of fossil fishes coming from rocks suspected of being southern equivalents to the Aztec Siltstone further north was discovered here a few years before by a young PhD student named Ken Woolfe, now a lecturer at James Cook University in Townsville, Queensland. The fish fossils were only scraps of scales, teeth, and bones, but enough of a faunal assemblage was collected to prompt us into writing a small scientific paper outlining the importance of the fauna and recording the presence of Aztec Siltstone in the area.

The sun was shining and the winds were gusting at around 30 knots when we set off sledging across the blue ice that day. Patches of freshly fallen snow were pleasantly smooth to sledge over, but each time we hit the jagged surface of the solid blue ice of the glacier the sledges would lose control as the winds made them drift and sway behind the skidoos.

Stronger winds developed as the day wore on, causing the sledges to blow around even more on the ice, making them difficult to control. Powerful freak gusts would randomly jerk the skidoos and push the sledges close to toppling over. After a few hours of this we stopped to put the metal guide pins down through the sledge runners a notch or two deeper, so as to give more effective grip on the ice surface. It was hard work for the two people on back of the second sledges to steer them over the blue ice during these gusts. Our immediate fear was that the sledges would eventually overturn and crash around on the ice, causing damage to our gear or breakage to the sledge itself, so great care had to be exercised while crossing these wide fields of roughly hewn ice.

As the weather grew fouler I could feel my fingers and toes getting decidedly colder. It was only our first full day in the field, and so I didn't want to seem like a "wuss" and complain, so I said nothing. I was being dragged along behind the sledge in the strong winds and couldn't stop to adjust my clothing, so I had to just keep holding on tightly and working the sledge to try and keep warm. I looked over occasionally to see how Fraka was getting on. She looked rather cold and miserable. At the next stop we rummaged through our packs and donned more layers of clothing. I put on my bear paws. These are thick leather mittens with a fur patch on the back of the hands for wiping the streaming snot from your nose, a common condition that always develops when one is sledging head-on into strong cold winds.

At about 5:00 P.M. that day Fraka was looking seriously cold and couldn't seem to keep warm enough, so Brian and Margaret wrapped her up in her one-piece bunny suit over her already heavily packed layers of clothing. By this stage we had covered a fair distance from the base camp, about 20 kilometers across the glacier, and could see the mountains looming just up ahead. It was imperative to get into the sheltered bays near to the mountains as soon as possible where we might find some respite from the impending storm, so we pressed on relentlessly for another hour or so.

By 6:00 P.M. the storm was gusting fiercely, pushing our sledges around like toys on the glassy ice, even when we were not traveling. We reached the mountains but couldn't see any suitable campsite that could protect us from the roaring winds. Fraka, Margaret, and I huddled up behind the back of one sledge, trying to get a solid barrier between the biting wind and us.

Brian then heroically unhitched his Skidoo and went off alone towards the mountains, his vehicle being instantly silenced by the deafening roar of the blizzard. We crouched at the back of the sledges and waited for what seemed like an interminably long time, but in retrospect was probably more like ten or fifteen minutes, wondering whether Brian really knew what he was doing. The thought of whether or not we'd ever see him again did cross my mind.

Suddenly, much to our great relief, Brian appeared out of the white blowing snow in front of us. He flashed us a wicked toothy grin saying, in a deadpan voice, a line I will never forget: "If you want to live, come with me." Later I was to discover that this was almost the same line spoken by Arnold Schwarzenegger in the movie *Terminator 2*, when he grabs the heroine and drags her off to safety.

We hitched up the sledges and followed Brian into the storm, but about five minutes later we rounded a sharp rocky bluff and found ourselves in a wind shadow. It was still gusting, but nowhere near as fierce as out on the glacier from where we had just come. Out there we could see huge billows of white powdery snow streaming upwards in geyser-like clouds, changing directions frantically with each erratic windblast. People would not survive very long out in those conditions, I thought to myself.

The tents were hastily pitched one at a time with each of us holding down a corner. As soon as they were up we shoveled snow over the flaps, placed rocks on top, and then one person crawled inside to get the cooking gear set up. The others passed in the yellow kitchen box, the green radio box, the blue primus box and lastly the red food box was placed on the front tent flap close by the entrance. Both tents were soon set up in exactly the same arrangement. We covered up the skidoos with their nylon covers and secured them by anchoring them with tent pegs and guy ropes. Finally the last job to be done was to secure the twenty-meter-long radio aerial out on bamboo poles in a line perpendicular to Scott Base, with the two thin end wires pushed in through the ventilator pipes at the top of our tent to connect with the radio inside. All of us then quickly disappeared into the shelter of the tents to get warm.

The comforting roar of the primus stoves quickly heated up the pyramidal interior space of the tents. The first thing one did after getting inside the tent was to melt snow on the stove so that we could each have a hot drink. We found ourselves ravenously hungry after the day's hard traveling and setting up camp in the strong winds. It had been quite an ordeal for our first full day in the field. Could it get worse than this, I thought to myself, or was this considered to be just an average day in the deep field?

It was my turn to cook that night, following on from the order of cooking begun during our field training days. The pattern of our entire expedition was that each member of the party would take their turn to cook the meal and the others would come into their tent, when summoned, to eat together. That first night in the Cook Mountains started typically with some powdered soup and bread. The others chatted and watched as I then wearily prepared a packaged roast lamb for the main course. My notebook says that I boiled it in the plastic, only later discovering that one generally removes the outer plastic covering before cooking such roasts. Nonetheless, once the heat had thawed the thing we demolished it with gusto, along with some rice I had boiled and smothered in the morning's leftover bacon fat (still frozen to the fry pan), and some assorted frozen vegetables. To finish off the day we each had two small nips of Drambuie. I cleaned up our dishes with paper toweling, and we all turned in for the night to our respective tents. Outside it was hazily white and snowy, with gusting fierce winds up to 70 knots. Not nice at all.

Storms in Antarctica seemed to coincide with slightly warmer temperatures because stagnant cold air sitting on the ice is pushed aside by the rapidly descending katabatic winds. These are winds that have picked up speed by gravitational force from sliding down from the elevated polar plateau. The local topography of each region can greatly influence the nature of these winds as they flow down from the polar plateau. Mawson's base on Cape Denison was a particularly windy location that registered a mean wind speed of 46 knots (24.9 meters per second) for the entire month of July 1913. Mawson starts his chapter entitled "The Blizzard" with this observation: "The climate proved to be little more than one continuous blizzard the year round; a hurricane of wind roaring for weeks together, pausing for breath only at odd hours."

Mawson later gives a powerfully descriptive summary of what it is like to actually go outside in a blizzard in Adelie Land, when man is exposed to the peak of Antarctic weather at its furious worst, in the midst of a cold dark winter:

> Shroud the infuriated elements in the darkness of a polar night, and the blizzard is presented in a severer aspect. A plunge into the writhing storm swirl stamps upon the senses an indelible and awful impression seldom equaled in the whole gamut of natural experience. The world is a void, grisly fierce and appalling. We stumble and struggle through the Stygian gloom; the merciless blast—an incubus of vengeance—stabs, buffets and freezes; the stinging drift blinds and chokes.

I stayed outside only a minute or so to brush my teeth that night before quickly clambering back in through the small round portal of the tent and throwing off my various outer layers of clothing. In only my long johns, I burrowed down inside the four layers that made up my sleeping bag: an inner cotton liner, a thick down inner sleeping bag, a thick outer down bag, and a heavy canvas sleeping bag cover. As I lay there all I could hear was the deafening roar of the tent sides flapping violently in the wind. I hoped that it wouldn't rip and expose us to the forces of nature outside. If it did we would all have to bundle into Margaret and Fraka's tent or put up our emergency small tent, although this wouldn't really be an option in such strong winds.

There's not much to say about the next day. Holed up in our tents waiting for the weather to clear, we couldn't do very much, although after the strain of the previous day we were all very glad to rest. We passed the time reading, making cups of tea, and occasionally taking short naps. Eventually as the boredom got to us we ventured out for short walks to the desolate rocky outcrops near the tents. These rocks, I soon discovered, were the basal non-fossiliferous layers of the Beacon Supergroup named the Junction Spur Sandstone. However, each time I braved the weather and went for a stroll to the rocks it soon became too windy, as my face and fingers got very cold from the high wind-chill factor.

None of us stayed out for more than about 20 minutes at any one time that day. In the afternoon we poured over our geological maps, read scientific papers on the geology of the Transantarctic Mountains and prepared ourselves for the first bout of fieldwork that we would tackle as soon as the bad weather cleared.

The assault on Gorgon's Head was the first aim of our expedition, but at that time we could only snatch glimpses of it through the snowy gusts. It was a distinct, pyramid-shaped mountain that towered skyward for some 1960 meters. The walk would be about seven kilometers as the skua flies from our base camp to the top of Gorgon's Head. It was one of several high peaks in the shadow of Mt. Longhurst, the highest point in the Cook Mountains at 2846 meters. It was not such a long walk, but because the weather could unexpectedly turn nasty on us we would have to take lots of extra provisions: climbing gear, crampons, ropes, a small portable tent, extra food, a primus stove, some fuel, and so on; in fact, everything required just in case we were holed up in another storm on top of the mountain and had to wait several days before we could return to the base camp. I could sense already in the planning stages that we would be quite loaded up even before setting out, but as our lives were dependent upon the degree of careful preparation, we agreed unanimously that all of these precautions were absolutely necessary.

After dinner that evening we all thought that the weather was about as bad as it could get. That evening the radio schedule from Scott Base indicated that the storm was still peaking and that more foul weather was due the next day. Later that night I read for a while but couldn't get to sleep very easily, mainly because we hadn't been active much that day. I snuggled down into my sleeping bags and pulled my balaclava down over my eyes to keep out the light, and eventually drifted off into a shallow slumber. I recall waking several times that night, and not being particularly sleepy, lay there in my bag, warm and cozy, thinking of my wife and kids back home and listening to the howling gales outside. The thought of being so isolated was starting to sink in; tiny bites of homesickness already gnawed at me, and I'd only been away from home for about three weeks.

The next morning annoying winds gusted between 30 and 40 knots, although it was a mild 5°F. The barometric pressure was still dropping—not a good sign at all. The storm seemed to have settled in for a while. Our morning radio schedule at 8:00 A.M. was marred by static as the sticky weather hindered clear communications. Somehow, though, we managed to send a garbled "We're OK" message back to Scott Base to allay their fears. I ate a light breakfast and nodded off to asleep again, sleepy from the previous night's restlessness. Apart from having to go outside twice to urinate, I otherwise spent the whole afternoon inside the tent, snuggled deep inside my sleeping bags, reading my novel, Eric Van Lustbader's *Shan*.

I went for another short walk over to the rocks that afternoon and gazed with hungry eyes at Gorgon's Head. I could almost feel my hands chipping away at the shale up there, imagining myself finding all sorts of amazing fossils, probably skulls of new fish species which were waiting to be collected from every rocky ledge! If only this damned weather would clear up, I kept thinking. It was still gusting very strong winds, blasting my face with large ice particles that melted on contact. I wandered back to the tent sullenly to find a cheerful Brian busily preparing dinner.

In my diary that day I recorded a short note about my clothing. Outside I was wearing long johns (thermalites) with polar fleece salopettes, a thick woolen shirt, woolen jumper, yellow one-piece bunny suit, wrist warmers, and my black leather sledging hat, with mukluks and thick woolen socks on my feet. This attire worked quite well for short forays from camp. I also noted how far we had gone on the skidoos: we actually covered 27 kilometers from our first campsite. I must have been bored out of my brain that day because I kept recording barometric pressures to pass the time, hoping the storm would change its course. The barometric pressure was 858 millibars at 5:07 P.M. and 855 millibars at 7:40 P.M., still dropping. Still not a good sign.

Brian prepared a somewhat unusual culinary feast that evening comprising a stew made up of macaroni, bacon, cheese, peas, and onions, all thrown in together in the camp oven. On the subject of food I am often asked did we eat well, and did we get bored with the prepackaged food boxes. The answer is yes, we ate very well, and no, we never got bored with the meals because we had experienced deep fielders like Margaret and Brian on our team. They brought an additional box of spices, herbs, and cooking items to enhance our regular supplies with a never-ending variety of spice or curry combinations. One can eat well using the dehydrated packets of lamb roast and turkey tettrazine, but by adding some rehydrated onions and a few dashes of curry powder, and making up a jazzy sauce using some of the spare packet soup mixes, or lemon-flavored drink crystals, one can quickly whip up a meal fit for a king or queen (of some obscure third world country). Whatever, we always ate well, and not once did any of us complain about any of the others' cooking. Our cooking skills certainly "evolved" as we got deeper into the journey. I have even included a couple of our more unusual recipes in an appendix at the back of this book.

Food plays a vital role in keeping up the spirits and morale of anyone working in the harsh Antarctic conditions. The standard NZARP food boxes that we used contained enough food for 20 man days (that is, five days for the four of us): three feeds of frozen meat, fish, or poultry (which were so large that each meal for two men was enough for the four of us to eat well); two 500 g blocks of cheese, two salamis, ten large (250 g) blocks of Cadbury's chocolate, two packets of dehydrated stew (to feed four people), several packets of dried soup mix, bacon, dehydrated vegetables like potato, peas, and onions, packets of frozen vegetables, boxes of oats or cereal, a slab of butter or tub of margarine, packets of sultanas, dried apples or apricots, a small tin of instant coffee, a packet of tea bags, a small container of honey, some salt, and sugar. In addition, one can add "extras" like some fresh bread, dried egg powder, flour, and spices, if you ask the Scott Base field store man nicely and beg the chef on bended knee.

Some of the earliest Antarctic expeditions showed that people could survive on the Antarctic islands with what little food is available through the abundant seals and penguins. On the Swedish Antarctic expedition of 1903, under the leadership of Dr. Otto Nordenskjöld, their ship the *Antarctic* was crushed in the sea ice and the men survived an epic winter of harsh endurance on Paulet Island. In Nordenskjöld's book the expedition's botanist, C.J. Skottsberg, is quoted:

> Still our dinners are not always plain ones. Saturday is the best day in the week, for the man who does not eat his fill then has only himself to blame. Dinner that day consists of an endless number of seal steaks, and a plate of what is alleged to be fruit-syrup soup. I shudder when I think of the portions we received: seven or eight enormous black steaks, swimming in fried train oil, and garnished with bits of blubber.

Mawson's men survived on the following sledging rations: 230 g pemmican (beef and fat mixture), 340 g plasmon biscuits, 28 g cocoa, 113 g sugar, 142 g dried milk, 57 g butter, and 7 g of tea. This 917 g of food was reconstituted with hot water to double in weight. Scott admitted in 1902 that on his first expedition he didn't provide enough food for the sledging parties, so he increased the amount for his second expedition to about 30 ounces of food per man per day. This was still to prove an inadequate diet both nutritionally and for the required calories needed to keep warm. At extreme temperatures like –40°F one can use half of the food eaten each day just to keep core body temperature up to normal and avoid hypothermia.

However, despite the monotony of this diet, they always referred to the enjoyment of a good hoosh—a mixture of pemmican, boiling water, and varying amounts of biscuit or other ingredients. On Scott's last dash to the pole the return journey was fraught with disaster. Food depots were sometimes hard to relocate, and days on end of cut rations made the men's thoughts always drift back to food. Scott writes in his diary on 28 January 1912, on their return journey from the pole:

> We are getting hungrier. The lunch meal is beginning to seem inadequate. We are pretty thin, especially Evans, but none of us are feeling worked out. I doubt if we could drag heavy loads, but we can keep going well with our light ones. We talk of food a good deal more, and shall be glad to open out on it.

Yet the food situation can be very serious when things go horribly wrong, such as on Mawson's Far Eastern Party's expedition. On 13 December Lt. Ninnis disappeared down a deep crevasse with the sledge that contained most of the party's food supply. The two remaining expeditioners, Douglas Mawson and Xavier Mertz, had about ten days of food for themselves and nothing for the six dogs that were pulling their sledge. They were about 506 kilometers from the hut and in unpredictably bad weather conditions. Eventually, as they ate the last of the food, they were forced into a regular pattern of killing the weakest dog to feed the other dogs and to provide meat for themselves, doing this at ten-day intervals. Mawson describes how on 28 December they shared a meal of the last sledging dog, Ginger:

> As we worked on a system which aimed at using up the bony parts of the carcass first, it happened that Ginger's skull figured as the dish for the next meal. As there was no instrument capable of dividing it, the skull was boiled whole and a line drawn round it marking it into right and left halves. These were drawn for in the old and well established sledging practice of "shut-eye," after which, passing the skull from one to the other, we took turns about in eating our respective shares. The brain was certainly the most appreciated and nutritious section, Mertz, I remember well, remarking specially upon it.

This last comment is particularly disturbing in view of what was to come. Mertz died on 8 January 1913 from eating too much dog meat (specifically the liver), a condition known as hypervitaminosis (or, more specifically, vitamin A poisoning). Later as Mawson struggled on against impossible odds, alone, with little food, he writes about his last remaining food rations on 17 January 1913: "The day's march was an extremely heavy five miles; so before turning in I treated myself to an extra supper of jelly soup made from dog sinews. I thought at the time that the acute enjoyment of eating compensated in some measure for the sufferings of starvation."

Finally, after surviving a crevasse fall that drew on his last reserves of strength to haul himself out, he found a food bag left at a cairn on 28 January 1913. The effect the food had on his body and spirit is apparent from his words: "Hauling down the bag of food I tore it open in the lee of the cairn and in my greed scattered the contents about on the ground. Having partaken heartily of frozen pemmican, I stuffed my pocket, bundled the rest into a bag on the sledge, and started off in high glee, stimulated in body and mind."

I felt fortunate that we modern expeditioners were supplied with good quality food for the whole of our fieldwork. We never had to think about food much, just reach into a red food box and grab something to eat or cook. It's only when you run low on supplies that you start to miss luxury items, like coffee or chocolate. We got by comfortably with our supplies and didn't have to resort to eating our means of transport as Mawson and Mertz were forced to do.

Somehow, I don't think stewed Skidoo would be as nice as dog soup anyhow.

13
Dancing on the Gorgon's Head

For this far violet line could be nothing else than the terrible mountains of the forbidden land—highest of the earth's peaks and focus of the earth's evil; harborers of nameless horrors and Archaean secrets. . . .

—H.P. Lovecraft

I woke up excitedly next morning at 6:30 A.M., quickly poked my head outside the tent and was relieved to see that the weather had improved. I was brimming with enthusiasm over the prospect of getting out onto the mountain to collect fossils. The winds had dropped to 20 knots and the temperature a cool, but tolerable 1°F. We decided that it was suitable weather for us to ascend Gorgon's Head. After a very hearty breakfast we made our scheduled morning radio call to Scott Base, prepared our packs with the necessary gear and food, and set off from base camp at 9:22 A.M. We drove the skidoos about three kilometers from the camp to a low rocky spur that led directly up to the top of the mountain.

It was a long, hard slog to the top, which took us around five hours, mainly because we were all heavily loaded up with the extra survival equipment and food supplies. The mere thought of finally getting up to some fossiliferous layers had put me in high spirits, along with the rest of the party who were also keen to see some geological action up on the top of the mountain.

On the climb up I collected some fossil plants in the scree fallen from the steep rocky sandstone bluff just below the distinctly colored Aztec Siltstone layers. On my return to Australia I later showed these specimens to fossil plant expert Steven McLoughlin, then at the University of Western Australia, and we wrote up a description of the fossil plants, which was published in the British *Geological Magazine.* The most distinguishing features on the lycopod fossils were the different shapes of the leaf scars on the stems. Lycopods don't have true "leaves" in the botanical sense; instead they have outgrowths of the bark or stem as leaf-like structures. These eventually detach from the stem as they grow outwards, leaving a characteristic scar shape that can be like a diamond, rectangular or hexagonal and have a variety of distinctive inner markings as well. To an expert like Steve, the few fossils I'd managed to collect that day were a wealth of new scientific information.

The specimens of horsetails (lycopods) were identified by Steve as *Haplostigma lineare*, a species otherwise only known from the Late Devonian of Australia; *Malanzania*, a genus known previously only from the Carboniferous deposits of Argentina; and a species of *Archaeosigillaria* previously known only from the Devonian deposits of Africa and South America. All of these fossils smacked of "Gondwana," reinforcing the important biogeographic links that fossils give for linking the modern-day southern continents.

Staring at the squashed impressions of 380 million-year-old plants while on the top of an Antarctic mountain on a gloomy overcast day brought to mind a very different picture of what this region was like at the time. The land had just been colonized by life for the first time in the Silurian Period, about 420 million years ago. At the time these Antarctic fossil plants were thriving, there would have been a sparse forest of low tree-like plants (the lycophytes) and an extensive ground cover of small stem-like plants (called proteridophytes). These would have flourished near the waterways. The largest of these primitive "trees" were giant lycopods reaching upwards of 20 meters, but most of the plants grew to less than a meter above the ground. No colorful flowers or pendulous pine cones here, just featureless thin green plants having simple flattened extensions of their stems for leaves.

A few primitive invertebrates crawled around these forests, such as springtails, mites, scorpions, millipedes, and centipede-like animals, as well as early spiders and their close cousins, the trigonotarbids. These all fed largely on each other as herbivory hadn't yet been invented. Creatures just didn't have the specialized ability to digest cellulose in those days so they just had to be content to eat each other. On the land the scene was essentially quiet, but the real action in this Devonian world was taking place below the water, in the large flowing rivers and expansive lakes that teemed with many kinds of primitive fishes. Some were the size of tiny minnows; others were huge predatory monsters four meters long with razor-sharp stabbing teeth six centimeters in length. This was the real reason I had come to Antarctica, to collect the fossil remains of these most interesting ancient fishes from a time when fishes were the highest form of life on Earth.

We reached the base of the Aztec Siltstone, near the summit, at around ten minutes past three that afternoon. Finally I was face to face with the variegated green, red, and grey-banded rock unit famous for its ancient fossil fish remains! After about three hours of searching I had located a fish fossil horizon about twenty meters from the top of the Aztec Siltstone, more or less at the same level where the original fossil fish material had been found by Ken Woolfe's party three years earlier. The fossil bones showed up as light specks and streaks in a dark green, silty sandstone. These layers of rock did not outcrop very extensively and were spasmodically covered by the loose scree. I could only trace the fossil-bearing layers as small lenses of about two or three meters each in length. Further up the hill near the top of the Gorgon's Head we located another two fish-bearing horizons and proceeded to spend an enjoyable afternoon collecting lots of specimens. Most of these were small fragments of bone or teeth, nothing of spectacular appearance to the layperson, but many represented new records or unknown species for this region, so I was extremely happy with the things we were finding. Often this work required sitting down with a pile of rocks and carefully examining each surface with a hand lens to find the really interesting small things, like the beautifully preserved teeth of sharks or exquisite fish scales adorned with complex sculpturing on their surface.

As I worked on hands and knees with my nose down on the rocks, Margaret and Fraka measured a detailed section through the Aztec Siltstone, aided by Brian who held the measuring staff and lugged gear around for us. The sky was an ominous dark grey, reminding me of the gloomy Melbourne winters of my childhood but, as there was no wind or snow, we were able to keep working up on the top of the mountain until about 8:00 P.M. Then, hitching our full backpacks bursting with specimens, we turned our back on Gorgon's Head and started heading back to camp.

It was a little easier going downhill, although we were careful not to move too fast through the loose scree slopes as our backpacks now added a lot of extra weight to each step. The last leg of the homeward trek involved donning crampons over our boots and leaping over a series of small crevasses in the slippery hard ice.

I have vivid memories of that day for it was the first time I'd worn my double leather climbing boots which I'd hired at the rate of $NZ13 per week. My feet were in agony by the end of the day as the boots were not a comfortable fit. I never wore those blasted boots again after that day, instead switching to the lighter Sorrell boots, or mukluks if not venturing far from camp. We reached our base camp, somewhat sore of foot and quite exhausted, yet elated with the good day's work, just before midnight.

It was Margaret's turn to cook dinner. Being tired we decided that one of the dehydrated meals would suffice, served with whipped potatoes lashed with butter and garlic powder. We jovially washed down the meal with a few nips of Irish Cream. I recall that Margaret had the "clumsies" that evening. She knocked over the water pot, then her drink, making little mistakes because she was so tired, as we all were. After cleaning up we all turned in to our bags around 2:00 A.M. Brian then tried to radio Scott Base but we received no reply. We must have fallen asleep close to 2:15 A.M., after twenty hours of activity since we awoke.

Next morning I woke up close to noon after a very deep sleep. It was slightly warmer at 14°F, but overhead the sky was still overcast. Light winds were blowing. My body ached all over from the previous day's massive effort; I sported blisters on my feet and had sore ankles from the offending leather climbing boots. After lunch we set out around the bend on skidoos to check out a low bluff of greenish-grey sandstone that Margaret was interested in examining for trace fossils, the evidence in the rocks of where life had once been moving around. We cruised around the face of the mountains on the skidoos for about six kilometers to reach the outcrop that comprised about 60 meters of exposed lower rocks of the Beacon Group. Margaret told us it was probably the Hatherton Sandstone, named after Trove Hatherton, a geological pioneer of the New Zealand Antarctic Program.

These rocks were much older than the rocks with the fish and plants that we had been collecting the day before. They were of uncertain age, but Margaret suggested that they were probably early Devonian, maybe 400 million years old and, more importantly, that they represented a completely different ancient environment. These rocks, unlike the ones on top of Gorgon's Head, were full of trace fossils, the burrows, and feeding trails of invertebrate animals. Margaret busied herself that afternoon taking many measurements, photographs, and samples while the rest of us looked around trying to spot any unusual fossils. We found several nice examples of a large burrow form known as *Beaconites barretti*. The burrow was about as thick as your arm and had fine curved layers in it, indicating that the animal, probably some sort of crustacean, had been pushing down layers of sand with its legs after each episode of digging.

That night I was perhaps a tad ambitious in the culinary department. I cooked Thai fish with dried whole chilies, coriander, lemon grass, soy, and pepper; and served it with fluffy white rice, stir-fried veggies, and mushroom soup. It wasn't bad, despite having to use powdered condiments and spices. My one major complaint to the New Zealand Antarctic Program is that you just can't get fresh chilies, coriander, and basil when you need them! We had one nip each of Stone's green ginger wine with the meal. Everyone applauded the meal and we even saved the leftover sauce for another time. Around 11:00 P.M. we all retired to our sleeping bags with the plan to climb another high peak the next day if the good weather held out.

The temperature was 7°F when we awoke next day. Weak winds blew from the west, and the sky was still quite overcast with some scant blue patches. This turned out to be another full-on workday. In the morning we set off once more on the skidoos and drove back to the rocky outcrops where we had worked the day before. After first scanning the hills through the binoculars we decided that we could see no signs of the Aztec Siltstone on the neighboring peak, nor an easy access route to the top, so we then changed plans for the day. We drove further down around the base of the mountains to some low craggy outcrops of the Junction Spur Sandstone. It was an interesting section to look at as it showed the transition from this basal unit to the overlying Hatherton Sandstone.

Margaret was particularly interested in searching for more trace fossils, so she and Fraka did a detailed measured section of that outcrop and recorded the relative densities and abundances for all the different species of trace fossils. This work involved actually measuring the thickness of each individual layer of uniform rock in centimeters, noting precise details of its composition, structure, and presence or absence of trace fossils. The only fossils I recall seeing that day were masses of *Skolithos*, a perpendicular pipe-like burrow in the rock, and more of our regular layered burrows called *Beaconites*. Both of these fossil burrows were made by varieties of long-gone arthropods, most probably a crustacean of some sort. Our reasons for believing this relate to what happened later in the trip. At Mt. Gudmundson on 3 December Margaret discovered a resting trace of the animal that probably made the larger *Beaconites* burrows. She studied that resting trace intensely and took several photographs of it, as it is the only evidence we have that *Beaconites* burrows were made by some sort of arthropod. If my memory serves me well I recall that it was quite possibly something like a horseshoe crab.

In my notebook for that day there is a drawing with a cryptic figure below it saying 750-800 per square meter, giving the density for the number of burrows of *Skolithos* in the rocky layers. There is also a drawing showing the two kinds of *Beaconites*; one shows the larger *Beaconites barretti* (like a fat salami structure in the rock) and the other was the narrower *Beaconites antarcticus* (looking more like a thin cabanossi sausage structure).

It is interesting to note that this trace fossil, *Beaconites*, now known from various localities around the world, was first described and named by an Estonian geologist, Professor O. Vialov. It was based on his study of photographs of various trace fossils and sedimentary structures taken at Beacon Heights by geologist Larry Harrington of the University of New England (Australia). (The name *Beaconites* is obviously derived from the Beacon Heights mountain, first named by Scott on his 1901 expedition because it could be seen from many miles away.) *Beaconites barretti* was named in honor of geologist Peter Barrett of Victoria University in Wellington, a veteran of many Antarctic expeditions covering most of the Transantarctic Mountains. *Beaconites antarcticus* was named by Margaret Bradshaw in her excellent paper on Antarctic trace fossils published in 1981.

There is still some controversy raging over the nature of these trace fossils and what ancient environments they could possibly represent. Margaret argued strongly for a shallow to near-shore marine environment for the trace fossil assemblages, whereas Ken Woolfe, a Victoria University geology graduate and veteran of several Antarctic expeditions himself, published an opposing view in 1990 that these rocks were mostly freshwater, formed in or near ancient river systems. Nigel Trewin, a Scottish geologist, and Ken McNamara of the Western Australian Museum have both studied the trace fossils found near Kalbarri in Western Australia, a very similar assemblage to what occurs in the lower Beacon Group sediments of Antarctica. They go a little on both sides suggesting that the Antarctic assemblages, if formed in similar conditions as the Kalbarri trace fossils, were probably formed close to marine conditions, with the influence of both wind-blown sands (forming coastal sand dunes) and freshwater sandy delta deposits.

Friday, 22 November, marked our seventh day in the field. We awoke to –1°F, with moderately strong winds gusting around thirty knots and a barometric pressure of 873 millibars and still falling. We heard from the radio schedule that we would be re-supplied by the Twin Otter aircraft between 27 and 29 November. This raised the issue of what things we would need on the re-supply and our list included color film, mattress rucksack buckles, two lashing ropes and two cargo straps plus loads of assorted batteries (mainly for our walkmans). In addition we would require another six drums of Mogas for the skidoos and one drum of kerosene for the stoves. We had plenty of food but requested a small bag of coconut and some peanut butter in order to attempt some exotic dishes further down the track.

The immediate plan was to pack up and tackle the McCleary Glacier, but we were still a little concerned that the wind could pick up and turn nasty, so we decided to wait and review the situation later that morning. Unfortunately, the wind never eased up, so we stayed in our tents reading and napping most of the day. I noted in my diary that I felt quite tired for most of that day, and I added a peculiar note that "strangely have had to urinate frequently." In this case it didn't develop into anything serious. It was more likely that I had been drinking too much tea and coffee all day. After a dinner of dehydrated beef stew curry, Fraka, Brian, and I played cards for a while, while Margaret braved the outside wind to pack her trace fossils.

My field diary also records "tonight we began reading aloud *At the Mountains of Madness*." My spirited reading of the first few pages, setting the mood of the book, transfixed the others. They all thought it was an excellent idea to bring a book based around an Antarctic fossil-hunting expedition and read a few pages out aloud each night.

I could sense we were all eager to start moving up the McCleary Glacier, thereby entering an unknown region of Antarctica, a place on Earth where no humans had ever ventured before. The very thought made my head spin, thinking of how the first polar explorers went boldly into new territory, but often with disastrous results. However, I reassured myself with the facts that we were equipped with the best modern equipment, excellent maps, and aerial photographs, and had personnel with a total collective Antarctic experience of more than ten seasons of deep field expeditions.

I slept well that night, despite the howling gales outside. I knew we were ready for anything.

14
Up the McCleary Glacier

The silence was deep with breath like sleep
As our sledge runner slid on the snow
And the fate-full fall of our fur clad feet
Struck mute like a silent blow
On a questioning "hush" as the settling crust
Shrank shivering over the floe;
And the sledge in its track sent a whisper back
Which was lost in a white fog-bow.
—From "Barrier Silence" by E. Wilson, 1911

Edward Wilson's poem beautifully portrays the feel of sledging, although I must point out that he wrote it with particular reference to their more rigorous routine of man-hauling the sledges. The next part of our trip involved a full day's sledging, and I recall the memory of this day with vivid clarity whenever I read Wilson's poem.

We were attempting to become the first team to cross the Cook Mountains from the Darwin Glacier to the Mulock Glacier. In actual fact, we were not the first field party to visit the northern part of the Cook Mountains. New Zealanders Harry Ayres and Roy Carlyon visited this region from the Polar Plateau side during Christmas 1957. After climbing Mt. Ayres they crossed Festive Plateau eastwards (so named because they were there at Christmas) and set up a survey point near Mt. Longhurst, which unfortunately couldn't be used because of cloud cover. Ayres and Carlyon had an epic trip. Ayre's dog team fell into a crevasse and it was a real struggle to get the dogs and the sledge out, yet surprisingly only one dog was killed. Ayres went right to the bottom of the crevasse to retrieve the Christmas pudding his wife had made!

Covering 50 kilometers by sledging up a glacier route where no one has ever been before is a good day's work in anyone's books, so we were very pleased at the end of that long day. The weather remained mercifully clear with only a little wind. We made our camp on blue ice in fine weather at the end of about nine hours' traveling. My mind has a clear mental picture of the little yellow pyramid tents as colorful specks on a vast ocean of shining blue ice in the shadow of Mt. Hughes, a snapshot image viewed from climbing up a nearby rocky peak. It was the only time on the whole journey that we ventured close to the open polar plateau, where no mountains stood between the South Pole and us.

We packed up camp next morning in 1°F and headed down towards the Darwin Glacier, taking a very wide course around the protruding Tentacle Ridge to avoid a large crevasse field which was clearly marked on the map. Our course then followed straight up the middle of the Darwin Glacier towards it origin, then turned northwards up into the narrower McCleary Glacier. By about 6:00 P.M. we were traveling parallel with the mountain range to our right and with the open flat expanse of the polar plateau on our left.

We moved camp only about fifteen kilometers that day as the skua flies, but in order to get safely around the southern Cook Mountains we had to take an enormous detour. The Darwin Glacier was hard, jagged blue ice with no snow cover for most of the journey. After the first half of the McCleary Glacier the surface rises steeply from an elevation of 1000 meters near the southern end of the mountains to 1600 meters at the place we finally camped. My line drawn on the map shows us heading straight for a very large crevasse field, but we avoided it by coming in close to the mountains to camp that night in the shelter of two jutting ridges. These are unnamed twin peaks on the map, 1940 meters and 1970 meters high, respectively.

This was the easy part of the McCleary, a very straightforward day of traveling with no sledge overturns or any hidden dangers. The following day would be more challenging as the glacier rose up from our camp at 1600 meters to the Festive Plateau at 2200 meters, with a wide circle of large crevasse fields and ice falls all around it.

Margaret cooked a chicken for dinner that night and this was followed by another communal reading from *At the Mountains of Mad-*

ness. It seemed that everyone was enjoying the act of reading aloud to the rest of the group, using much theatrical license to accentuate the mysterious and unknown elements of the story.

The next morning heralded another epic day's traveling. Fine clear weather prevailed with little to no wind, making an ideal day for traveling, even though it was much colder at –6°F. I remembered from the recon and put-in flights that from the air the McCleary Glacier had some large crevasse fields, but somehow we managed to work out a narrow sledging route around these. Strangely enough when you are on the ground it's not quite the same, as a lot of luck and careful navigation is needed or you can easily find yourself surrounded by crevasses, or down one. As our destination that day was a potential new fossil site, situated a long way from our camp, there was no point stopping somewhere near the end. We were committed to go the whole distance so as not to have to waste another day packing up camp to move just a short distance further on. Setting up and packing camp could cost us almost half a day, so we had every reason to keep pushing onwards while the weather was good. This set a cracking pace for what proved to be a very long day of sledging.

We started off after packing up camp around 10:00 A.M., heading straight up the middle of the river of ice. At the head of the glacier we could see the ice rising up dramatically where ice falls appeared around us at different places. These are spectacular sights of glistening, rippled ice cascading like a frozen waterfall over a sharp drop.

The slope at the top of the glacier was fairly steep. The leading Skidoo didn't have enough gusto to pull up its two fully laden sledges, so we decided to take each sledge up one at a time. The skidoos still couldn't manage the load. Brian suggested we rope each of the four sledges to both skidoos and pull them up in tandem. This demanded much concentration as the two skidoos had to pull evenly and smoothly at the same speed, but it worked, so we eventually ended up with all four sledges at the top of the glacier, the edge of Festive Plateau.

We had a stunning view from there that words cannot fully describe. There was quite a distance to cover across the vast expanse of the mighty Mulock Glacier and, far away in the ethereal distance, soar-

The Wright Valley looking toward Lake Vanda. The Onyx River flows down the middle of the valley during summer. The flat-topped Dais can be seen behind the frozen lake as clouds surround the brooding snow-capped peaks of the Asgaard Ranges.

There were no motorbikes in the Devonian Period! These markings are giant fossil arthropod tracks discovered near Escalade Peak. At almost a meter wide, it was probably made by a huge sea scorpion up to two meters in length, more than 400 million years ago.

View from our camp across Lake Victoria looking at the head of the Taylor Glacier. Our first geological foray was to Sponsor's Peak (not shown), situated to the left of the glacier.

My initiation into the Polar Plungers Club on 7 November 1991. Note the harness to save me from being swept under the ice by currents. The outside air temperature was actually much colder at -25°C than the sea water (only -1.8°C). Whilst submerged, I noticed dark shapes looming around me. Seconds after I was pulled out, a large Weddell Seal popped out of the same hole.

A timeless scene where one can almost sense the presence of the early explorers. This is the inside of Scott's hut at Cape Evans. Scott's chair and mug are seen at the end of the table. Ponting's darkroom is behind the mug and Wilson's laboratory is seen to the left.

Inside Shackleton's hut at Cape Royds. Food supplies, kept in freezing conditions since the hut was abandoned in 1909, are still good to this day.

Our camp out on the blue ice of the McLeary Glacier in early evening. The hazy snow-drifts are rolling in from the open polar plateau.

A lunch stop en route to Escalade Peak, visible in the background. The dark rocks at the top are igneous dolerites, forcefully intruded as hot magma into the sedimentary rocks of the Beacon Supergroup, shown here as the lighter-layered rocks.

Out in the Skelton Névé. In the morning we would fill our overall pockets with frozen salami, which our body heat would thaw by noon. This gave rise to the catch-cry of "Is that a salami in your pocket or are you just glad to see me?"

ing up through hanging clouds and coiling mist, were the black-banded mountains of the Warren Ranges. They were about 50 kilometers away yet they looked so gigantic and mystical; something you'd imagine would exist only in Valhalla. Beyond these lofty peaks we could see the omnipresent and almost eternal white expanse of the polar plateau with its windy snowdrifts fading off into the hazy grey-white horizon.

Jet-black doleritic rock forming most of the Warren Ranges was lightly draped by snow, the hardness and darkness of the ancient rock in stark contrast with the softness and whiteness of the newly fallen snow. These rocks, formed by molten hot lavas squeezed at great depths in ancient times, were forcibly extruded between the layered sandstones of the Beacon Supergroup sometime during the age of dinosaurs. The formation of these rocks to me represented the antithesis to everything that the white snow symbolized—new, soft and cold. The sheer scale of everything in view was unimaginable—a giant river of ice more than 50 kilometers wide, jagged mountains of black and white towering up 2.6 kilometers, and the vast, white oblivion of the almost never-ending polar plateau beyond.

To gaze upon such grand sights of nature can be almost a mystical experience. It made me ponder in awe at the forces of the Earth and the billions of years of evolution that had eventually led us, mere frail humans, to this forsaken continent. It made me proud to be a scientist, someone who can appreciate such a spectacular view for what it was, a natural sculpture forged through millions of years of the earth's dynamic processes, without feeling any desire to question the philosophical basis for its existence.

We had lunch in view of the Warren Ranges on the edge of Festive Plateau. No detailed geological investigations were ever carried out in these mountains around us. Up ahead in the distance we could see bands of colored siltstones, the layers that often contain fossils. The next part of our journey would take us down from the high Festive Plateau to a place called Fault Bluff, about another fifteen kilometers away. The next leg of the journey involved a long even slope with ice falls surrounding it.

Descending slowly, we managed to take each of the four sledges down the slope one at a time by heavy use of their sledge brakes. The brakes are pieces of wood with metal teeth that grip the ice when you step down on them. However, if the slopes get too steep even the weight of a human standing on the brake may not be enough to slow the sledge down. In these cases one has to go very carefully so as not to let the sledge get out of control. As this slope became steeper the sledges then became more difficult to control. At one point we accidentally overturned a sledge and a trickle of brown liquid oozed out of the sledge onto the white snow. I bent over and smelt the brown mess. One of our precious bottles of Kahlua had bit the dust.

A short time later Brian suddenly threw his hand in the air and shouted for all of us to halt and not move an inch! He yelled out that there were large crevasses all around us. We stood there, silently gazing at the snow-covered ground, wondering what the hell we should do next to get out of this predicament. Brian took control. He asked me to rope up to him, suggesting that we walk down the slope ahead to pace out a safe route for the sledges. After harnessing ourselves together with long ropes, Brian grabbed the two-long-long crevasse probe from the sledge and carefully began walking about twenty meters in front of me, prodding the ground with the probe as he advanced slowly forwards. I held tightly to the rope in case he fell, ready to anchor him to the ground with my ice pick. As I walked down the ice slope I took care to step in his footprints.

After marking a safe path down the ice slope, we retraced our steps back to the sledges and then slowly drove down, lowering the sledges in front of the skidoos one at a time until we were all safely at the bottom. It was around 6:00 P.M. by then yet our destination, Fault Bluff, still loomed a long way up ahead in the distance. Fortune smiled down upon us, though, as from here it was an easy drive, the skidoos pulling both sledges along at a fast pace over the flat powdery snow.

We reached Fault Bluff, a dark angular mountain with faulted dolerite rock, around 10:00 P.M. that night. After quickly setting up camp a short distance away from the rocky outcrops, one of us immediately began making dinner. It was close to midnight by the time we ate, and although dog-tired we were exulted by the success of the last two days and chatted cheerily over dinner.

We had successfully pioneered a route up the McCleary Glacier and had arrived at a completely new patch of mountains bursting with potential for new fossil discoveries. We all slept deeply that night, and I dreamed once more of finding ancient fishes, the last known inhabitants of this desolated part of the world.

15
A Room at the Fish Hotel

Study the past, if you would divine the future.
—Confucius

Paleontology is the study of things long dead and gone. To some this might seem somewhat morbid but, as the words of Confucius say, the key to understanding what the future might hold is a thorough knowledge of what has transpired in the past. On this next part of the journey we worked at Fault Bluff, marking the beginning of our scientific discoveries. We found many new fossil sites loaded with prize specimens there. The high of discovery was about to override my adrenaline buzz from the last few days of sledging.

Our tenth day in the deep field was 25 November. It was a very cold morning when we awoke, the temperature plummeting to an all-time low of –17°F. I slept in till about 10:30 A.M., then we made a late morning radio call to Scott Base to tell them what things we needed for our re-supply, including a "wide-mouthed plastic drinking bottle" for myself.

Early that afternoon I walked over to the outcrop and climbed up onto the flat pavements of exposed rock. It was the first time on the whole expedition that we had reached a completely new locality containing the right rock types in which to find fossil fishes. The rock appeared to be like the typical greenish-red Aztec Siltstone we had seen a few days before on the top of Gorgon's Head. Within a few minutes of hunting I had found a layer bristling with fossil fish bones. In fact these were exceedingly well-preserved fossils containing visible bone cell spaces. Eureka! I got busily to work with my hammer and chisel, slowly extracting the specimens, armored plates of placoderm fishes. One specimen was most unusual because I could not readily recognize what species it was from, and this made me excited.

Placoderm fishes were the most successful group of backboned animals on this Earth for nearly 70 million years, becoming suddenly extinct 355 million years ago at the close of the Devonian period. Rather shark-like in overall appearance, they were characterized by having a mosaic of overlapping bony plates forming an armored shield to the head and front of the body. Although most placoderms were small fishes under a meter in length, near the end of the Devonian period some real monsters evolved, perhaps reaching sizes of six meters or more in length. The nice thing about placoderms for us paleontologists is that their bony plates made excellent fossils, and are easily identified due to their variable shapes, types of surface ornamentation, like tubercles or linear ridges, and several kinds of sensory-line grooves cut into the bone. I had started my paleontological career by studying placoderms and was by then quite familiar with the common species that should occur in these rocks. Yet here, on the first day at a new site, I was chiseling out a large fossil fish plate that I could not immediately recognize. Later, back home in the lab, I would identify it as being the median dorsal plate from the center of the back belonging to a moderately sized Beaconites, or joint-necked placoderm, perhaps a species about 50 centimeters in length. Unfortunately, the specimen was too incomplete to make a more detailed identification, except that I am sure it's not one of the fossil species previously known from Antarctica.

Margaret and Fraka began working on measuring the section layer by layer while I flitted off to search every protruding rocky ledge for fish fossils. I found a few more fossil-bearing layers, some of which had faint impressions of placoderm plates. One of them was a phyllolepid plate. Phyllolepids (meaning "leaf scales") were flattened placoderm fishes which had distinctive bony plates bearing narrow ridges of wavy, concentric ornamentation, making them easy to identify even from small fragments.

The significance of phyllolepid plates at this site was that the Aztec Siltstone exposed at this locality must represent the top of the succession, in the youngest part of the sequence, as phyllolepids were only recorded from the top few meters of Aztec in the Skelton Névé area to the north. Phyllolepid placoderms were a group that I was very familiar with as I had first worked on them during my doctoral studies. In 1984, I published a description of the world's best-preserved phyllolepid placoderms from the Mt. Howitt site in central Victoria. Previously the group had been known from only one genus, *Phyllolepis*, based upon only one articulated example found in Scotland. Isolated and fragmentary plates of phyllolepids were well known from the Late Devonian freshwater deposits of Europe, Greenland, and North America, but the relationships of the group to other placoderms had been often debated amongst fish paleontologists. When they were first described they were thought to be jawless fishes, and only in the 1920s had they been correctly assigned to the jawed fishes, specifically to the armored placoderms. Although phyllolepid plates had been recognized in Australia by Professor Edwin Sherbon Hills of Melbourne University since the 1930s, the new material I had studied from Mt. Howitt showed for the first time the complete preservation of the entire fish with superb details of their jaws, tail, pelvic girdles, and even the otoliths (ear stones) which filled the saccular cavities of the inner ear. From my descriptions I envisaged these flattened predators lying in wait on the lake or river bottom, slightly covered by mud or sand, until some unsuspecting small fish ventured above them. Then, like the lightning fast angel sharks of today, they would spring forth and grab their prey. Finding phyllolepids in Antarctica was very exciting because I was hoping to recognize the same new genus I'd named earlier from Victoria, *Austrophyllolepis*, to prove links between the Australian and Antarctic species. Only collecting more material would resolve this problem.

And indeed I did collect some very good material that afternoon. One specimen was a huge fish fin-spine stuck right in the middle of a large sandstone slab. I simply had to have it as it was the genus of fish called *Gyracanthides* that characterized the Gondwanan deposits of this age. Brian volunteered to chisel it out for me while I went off searching for other fossil-rich layers. It took him almost two hours to complete the job. When he'd finished the metal chisel was duly cold-forged into a weird, blunted shape. The effect of the extreme cold on our tools meant that we continually had to sharpen our geological hammers and chisels throughout the expedition.

That night was a particularly cold evening, the temperature down to –8°F at 8:00 P.M., so we decided to stay inside and play "pass the pigs." This did not involve eating bacon in large quantities, but was centered on the various ways two cute little plastic pigs could land when thrown into the air. It never ceased to amuse us and make us laugh at their peculiar positions. There was absolutely no skill involved, just the luck of the throw. To heighten our enjoyment of the game Margaret sometimes played her tape of Michael Jackson's *Thriller* album on the little Walkman suspended in the net at the apex of the pyramid tent. The only other music we had to listen to for the whole trip was some of Fraka's tapes, including Kenny G and Graceland. The tapes I heard the most were naturally Brian's, as we shared a tent together. Brian's music consisted of Joe Sastriani's *Surfing with the Alien* and *Flying in a Blue Dream*, Big Pig's *Bonk* and the Pogues' *Rum, Sodomy and the Lash.* The latter has a wonderfully earthy version of Eric Bogle's classic Aussie song "And the Band Played Waltzing Matilda," which I learned by heart and would often sing to myself while out on the mountain tops searching for fossils.

The next day we climbed up onto the southern face of Fault Bluff and continued our search for fossils. I traced the lowest fish-fossil layer of rock laterally across from the previous day's site but only found a few scattered bone pieces as impressions in the rock. The fauna from here contained the commonest of the armored placoderm fishes, two genera called *Bothriolepis* and *Groenlandaspis*, and scales of an extinct lobe-finned fish.

Bothriolepis was a strange-looking little fish that had most of its body and head enclosed in a box-like armor of overlapping bony plates, with two segmented arms coming out of its shoulders. These jointed appendages were somewhat akin to the segmented legs of a crayfish, and may have helped the fish push itself deeper into the muddy sediment on which it probably fed. Its little eyes and nostrils were situated close together on top of its head, making it look a bit like one of the armored South American catfishes (*Plecostomus*) that are commonly sold at pet shops. Despite its odd appearance, *Bothriolepis* was one of the most successful fishes to have ever lived on the Earth. Its fossil remains have been found on every continent, represented by over 100 distinct species, which lived over a time span of some 35 million years! I also have a soft spot for *Bothriolepis* as it was the first fossil fish I seriously worked on, having studied the Mt. Howitt *Bothriolepis* species for my honors year thesis in 1980.

Groenlandaspis is another placoderm fish worthy of mention. It was first discovered in the Devonian rocks of East Greenland in the 1930s, hence its name meaning "shield from Greenland." Alex Ritchie first identified it from Antarctica back in 1975, and since then he discovered it in Ireland, North America, Australia, and the Middle East, and I had recognized it in South Africa. It was also covered in bony overlapping plates, like all placoderms, but had weak jaws and a peculiar high-crested bone, somewhat like a dorsal fin, straddled its back. Its small rounded tubercles make it easily distinguishable from *Bothriolepis*. Both *Groenlandaspis* and *Bothriolepis* are very useful for age determinations of Devonian rock successions between regions. The detailed descriptions of the several *Bothriolepis* species from Antarctica published by Gavin Young in 1988 were used to identify the different levels (or concurrent time zones) within the Aztec Siltstone outcrops. The close evolutionary relationships of the *Bothriolepis* species occurring in Antarctica and Australia also provided further support for the close proximity of these two continents back in the Devonian period, and their evolutionary radiations give clues to the possible dispersal routes for some of the *Bothriolepis* lineages.

In the upper units of the Aztec Siltstone section that afternoon I found some interesting lycopod plant fossils with more broken fish plates. Overall the fauna contained mostly placoderms like *Bothriolepis*, *Groenlandaspis*, phyllolepid plates, but also some scales, isolated bones and teeth of osteolepidids and rhizodontids (both extinct groups of large predatory bony fishes) and fin-spines of the acanthodian *Gyracanthides*.

Fraka Harmsen's skilled eyes could read the rocks as if she were browsing through a travel guide to some ancient faraway land. She readily identified abundant overbank deposits and classic sandstone point bar deposits, which told her that we were standing in part of what was once a large meandering river system about 380 million years ago. Most of the bones were situated in the lag beds of the old river channels, hence their abraded and fragmentary preservation. To find whole, complete fish fossils we would need to find ancient lake deposits where fine-grained mudstones and shale had slowly accumulated.

The only incident of note that day was that I almost stepped into a crevasse close to the rocky cliff when I was rushing to get onto the outcrop. I didn't think anything of it as I was jumping from ice to rock when my right foot went right through the ice and sunk down to my knee. I quickly pulled back on the other foot and was able to get a good footing on the rocks. I then leaped over the small crevasse, about a meter wide and very dark and deep, to some large rocks from which I could scramble up the cliff. I made sure not to walk over that icy stretch again during our stay at this locality. I didn't think much of the crevasse at the time, because I was too preoccupied with all the new fish finds to dwell upon small accidents.

Still, in my mind that night I went over the moment. If both feet had landed on that spot then instead of one foot breaking through the ice bridge I would have probably fallen straight down to the bottom of an icy tomb. I resolved to be more careful from that point on.

It was a freezing –8°F next morning and very windy. We packed up camp and moved around to the other side of Fault Bluff. It was rough going because the sastrugi, ice waves carved by the winds, were very high, making sledging a real chore. One sledge overturned, but no damage was caused. We had to move quite slowly not only because it was very windy but also because we were all starting to feel the cold. At around 3:30 P.M. we arrived at a low boat-like hill with an excellent exposure of sedimentary rocks. It looked like it had a lot of potential for fossils, so we decided to camp there. Camp was pitched by about 5:00 P.M. I walked carefully up to the outcrop and almost immediately started to find fish fossils scattered around in the loose rocks. At one locality large, complete fish spines and bony plates were found exceptionally well preserved in a concentrated bone-bed layer—a paleontologist's dream come true! Other silty fine layers had only small fragments of well-sorted bone mush. By the end of the day I had discovered at least six different layers with fossil fishes in them from the base to the top of the Aztec outcrop, so I knew this was going to be a very productive site.

Despite the temperature plummeting down to a chilly –15°F overnight, next morning it had warmed up to 3°F, and was a positively fine day with no wind. I noted from the topographic map that we were camped at an altitude of 1,900 meters. That morning, purely out of curiosity, I took some measurements of temperatures inside the polar tent with one primus roaring away. Surprisingly it was 32°F at the base of the tent near the ground sheet, 59°F at head height, and a sultry 77°F in the apex of the tent in the net where we hang our socks and gloves to dry. Once the primus is turned off the tent doesn't retain heat for very long, so we have to snuggle inside our sleeping bags to keep warm.

The day was spent collecting lots of great fish fossils. My notebook is full of notes and sketches detailing the new sites and showing the many kinds of fish fossils we found. I christened the site "Fish Hotel" because it was so rich with fossils that all the beds were full. At 3:30 P.M. I noted an important find in my diary: the ventral plate of a phyllolepid placoderm fish, the very same genus which I had earlier described from Mt. Howitt in Victoria, called *Austrophyllolepis.* It is the first record of this genus from Antarctica and a strong link with the fish faunas of this age in southeastern Australia. I photographed it and then spent quite some time trying to chisel the specimen out of the solid sandstone. Unfortunately, the rock split in the process of getting it out, although it wasn't difficult to restore the specimen when I got back home to Australia.

Next morning we talked on the radio to NZARP group K5076, led by Paul Fitzgerald, who were about to climb Mt. Adam in northern Victoria Land to explore its geology. They were the other NZARP deep field event that year so we shared a kindred spirit. Their mission was based on collecting granitic rocks for radiometric dating to determine the precise ages of the major geological events that have shaped the continent of Antarctica. To get the right samples they must sometimes scale high mountains, so it was no coincidence that most of their party was quite experienced at rock climbing. We kept in touch with them to chat occasionally, and enjoyed following their progress throughout our expedition. Occasionally we would relay messages back to Scott Base for them at times when their communications equipment was not working very well. In such cases they could get messages to us clearly, but not to Scott Base, which was actually much closer to them. We could not understand why this was the case, but one suggestion was that maybe when storms enveloped Ross Island, distorting communications to Scott Base, a clear atmospheric pathway may have remained open from the Darwin Glacier right through to northern Victoria Land.

One of the specimens I collected that day was very strange: an elongated median dorsal plate from a moderately large placoderm having very coarse, wart-like ornamentation all over its external surface. I suspect that it is probably a new genus of arthrodiran placoderm, quite unlike any previously found in the Australian-Antarctic fauna. It implies that not all elements of the fauna from Antarctica are common to the Australian assemblages, hinting that many more surprises could turn up with more collecting. Some other large arthrodire plates bearing similar coarse ornament were found later in the trip on Mt. Ritchie, giving me more tantalizing pieces of the puzzle. Working on placoderms is a bit like doing a jigsaw where you must discover the pieces hidden at different localities in widely different layers of rock. By recognizing the general picture on the puzzle (here being the ornamentation style on each of the placoderm plates) you can then end up with a few of the pieces, not necessarily the entire puzzle. Still, if you know your placoderms it might just be adequate to reconstruct enough of the picture to tell that it's something never before seen by human eyes. In this case, after several years of studying the fossils, I'm convinced it is a beast new to science.

I also suspected that day that there was another weird *Bothriolepis*-like placoderm in the fauna because of its different style of complex ornamentation. Like the other puzzling arthrodiran placoderm, I would need to find a few more distinctive plates of this species before its true identity might be revealed. Still, it was enough to get my curiosity roused and to know that I must search very carefully at the next few sites for more pieces to all these different puzzles.

Several kinds of large shark's teeth were collected that day which I tentatively identified in my notebook as *Xenacanthus*, which has distinctive double-pronged teeth, akin to two curved hooks rising up from a flat root. Although this genus of shark is better known from the younger Carboniferous and Permian deposits of the Northern Hemisphere, its lineage must have had Devonian ancestors somewhere, so why not Antarctica? I thought to myself. Yet instinctively I knew that these new teeth were not identical to the teeth of Northern Hemisphere *Xenacanthus*. The new specimens lacked the raised central "boss" on the root that characterizes typical *Xenacanthus* teeth. I had a hunch they would probably turn out to be another completely new genus of fossil shark and, eventually, they did.

As seen from these finds, it was a very exciting day with so many new discoveries, some new kinds, some unrecognizable species not previously found in the Aztec fauna and other puzzling and enigmatic species. Margaret and Fraka measured the outcrop that day at about 90 meters of exposed continuous section, with fish fossils occurring from right at the base to the very top few meters.

I kept working on the outcrop that day until 7:30 P.M. Tired and hungry, I then wandered back down to the campsite alone. We were all in a very good mood that night after our big discovery of the fossil sites and the excellent work on the geology of the region that Margaret and Fraka had accomplished.

Yet more fish fossils were collected the next day from the top section of Fish Hotel. I found well-preserved *Groenlandaspis* plates with many good *Bothriolepis* plates, lots of the large *Xenacanthus*-like shark teeth, plus some other unusual shark teeth that I had never seen before. About three of these weird teeth came from this site, which later would be described by Gavin Young and myself as a new genus. We named it in honor of the Australian National Antarctic Research Expeditions, calling it *Anareodus staitei*. The species name honors Brian Staite.

After lunch that day I spent quite some time carefully labeling, wrapping, and packing my specimens in some of the now empty food boxes. This system worked well as we could easily pack the boxes on the sledges, but it did limit our collecting due to space and weight restrictions. The way it worked was, the more food we ate, the more fossils we could collect!

After this chore was completed I cooked dinner that evening. We continued afterwards with our ritual reading of *At the Mountains of Madness*, and were then about halfway through the story. We were up to the unveiling of the ruins of the mysterious and ancient civilization found in the remote wilds of the Transantarctic Mountains, while at that very moment we were camped in a completely unexplored region of the continent. Time and time again, especially when wandering alone on a mountainside searching for fossils, I found my imagination wandering, thinking about sci-fi movies like Carpenter's *The Thing*, in which an alien spacecraft had crashed in Antarctica, its strange life form then taking control of one of the bases and killing all the humans. The fact that we were so isolated and alone there certainly heightened the thrill of our nightly readings from Lovecraft's masterpiece.

On Sunday, 1 December, we packed up camp and moved around to the next outcrop, another low flat-topped hill about five kilometers further north from Fault Bluff, towards the Mulock Glacier. A very good contact between the Beacon Heights Orthoquartzite and the underlying Hatherton Sandstone was exposed here. The Beacon Heights Orthoquartzite formed steep-sided solid cliffs 22.5 meters high, as we measured them using a rope dangled down from the top of the exposure. All up, the cliff was about 40 meters high to the start of the Aztec Siltstone layers forming its expansive flat top, which gave us an excellent view of our Fish Hotel hill in the distance. About 30-40 meters of Aztec Siltstone was exposed on the top and I found a few fragmentary fish fossils, some in very coarse pebbly conglomerate with individual rocks up to about twenty centimeters or so. My notebook is full of drawings of the specimens that we observed but didn't have time to collect as the matrix was very hard and it would have taken us too long. There was one very strange plate here, an extraordinarily long anterior median dorsal plate of possibly some unknown species of *Bothriolepis*-like placoderm. The rocks also contain large burrows of the *Beaconites* type; it is an unusual occurrence to find these in fish-bearing layers.

We finished working here around 5:00 P.M. and kept on sledging until we arrived at Seay Peak around 9:00 P.M., in the far north of the

Cook Mountains, facing the open Mulock Glacier. The Warren Ranges in the far distance were swathed in misty cloud rolling off the polar plateau, reminding me of some Tolkienesque scene out of Lord of the Rings. The weather was clear with a temperature of 5°F at 9:15 P.M. We had completed some 40 kilometers of sledging that day, as well as doing a lot of work at the outcrop five kilometers north of our previous campsite, so it had been a long but productive day.

The last few days had been the most exhilarating so far in terms of major fossil discoveries. We had collected many fine specimens from our new sites and Margaret and Fraka had measured detailed sections through the geology at each of the three new locations. The plan for the next few days was that we would leave most of our gear at Seay Peak, sledge over with one sledge each to the isolated hill just north of Kanak Peak, stay there one or two days, and then return back here to allow a full day to prepare for the coming of the helicopters on 7 December, our transfer across the mighty Mulock Glacier. If time permitted, we would also take a look at Mt. Gudmundson and possibly climb it for closer scrutiny.

We had completed sixteen days in the field and had achieved much. More importantly, our group dynamics was working well, we were efficient as a team, we all felt comfortable in each other's presence, and everyone pulled their weight when it came to the daily chores. Despite the ever-present pangs of homesickness that hit each of us from time to time, we were simply all too busy working each day to worry about much else. It felt like some crazy, wild, out of control roller coaster ride that I was on and now must see through to the end.

16
The Skua

Jonathan Seagull spent the rest of his days alone, but he flew way out beyond the Far Cliffs. His one sorrow was not solitude, it was that the other gulls refused to believe the glory of flight that awaited them; they refused to open their eyes and see.
—Richard Bach

I awoke on 2 December to find a clear fine morning, no wind to mention, and a bright sun beaming down on the snow-capped mountains encircling our camp. I snapped a photograph of Brian walking around outside in just his black long johns and Sorrells, the sunshine reflecting off his weather-beaten, well-tanned face.

Seay Peak is a pyramid-shaped mountain with black dolerite capping its apex. It peered down at us, casting its cold shadow across the horseshoe-shaped embayment where we were camped. Some of the multilayered Aztec Siltstone was visible in its face below the black volcanic rock, although as the front of the mountain formed almost vertical cliff faces it was difficult to reach the interesting fossil-bearing layers by any easy means. So, giving up on that idea, we decided to go for a short reconnaissance trip on the skidoos around the other side of Seay Peak to see if there were any accessible exposures of sedimentary rocks.

With two of us on each Skidoo, we playfully raced each other round to the other side of the mountain. Up the snow-covered slope we went, as far as we could reach before it became too steep for comfort. It was much windier, and colder, on that side of the hill because it faced the open polar plateau. It was soon apparent that no interesting rock layers were exposed here, only masses of black volcanic rock. We returned to camp, hitched one sledge to each Skidoo, then began making our way towards a little rocky hill just north of Kanak Peak.

The first part of the journey was slow going due to rough sastrugi and moderately strong winds rolling off the Mulock Glacier immediately north of us. However, once we rounded the spur, the rest of the journey became easy. There had been a recent snowfall so the skidoos and sledges glided smoothly over the new snow. The wind then eased off and changed direction, coming around behind us so as to make a thoroughly enjoyable ride to Kanak Peak. I recall getting the skidoos up to record speeds of about 20 kilometers per hour on one leg of this trip. We arrived close to 4:00 P.M., after stopping for a quick lunch somewhere along the way.

Lunch on these sledging trips is usually set aside before we leave and placed in compartments under the seats of the skidoos, which open up like motorcycle seats. Normally we would pack some crackers, cheese, salami, cereal bars, chocolate, nuts, and a thermos of hot orange or lemon drink. One neat trick we learned concerning lunch is to place one of the frozen salamis in the outside large pockets of our overall trousers so that by lunchtime the meat is partially thawed and ready to eat. This gave rise to the common lunchtime catch cry of "Is that a salami in your pocket, or are you just glad to see me?"

After reaching the campsite we erected our tents high up on the side of a slope because it was as close as we could get to the sandstone outcrops. In such cases it's just an easy matter of leveling out a small platform in the snow with the shovels to put the tents on. This position gave us a most magnificent view right across the valley to the south, a vista of many small isolated peaks of the Transantarctic Mountains stretching for many hundreds of kilometers, islands of black and cream rock in a sea of virgin white snow.

Later that evening I strolled up onto the rocky outcrops near the camp. They appeared to be of a kind of coarse-grained sandstone composition, probably one of the basal rock layers that contained many kinds of trace fossils, but unlikely to have any fossil fish remains. Nonetheless, I thought to myself that it would be interesting to search for fishes, almost convincing myself that maybe, with intense searching and wishing hard enough, I could probably still find something of scientific interest. Unfortunately, I did not.

The next morning (3 December) was again clear and sunny with only a slight breeze. Once again I pulled myself up onto the rocky ridge and soon had my head down searching intently for fossils. The others were over on some other part of the hill a fair distance away, so I was alone on my part of the rocky ridge. Suddenly a shadow passed quickly over the rocks, from above, scaring the living daylights out of me. We had not seen another living creature for almost three weeks now, so a moving shadow was a strange event. I looked up and immediately saw a bird, a brown skua gull, hovering delicately in the wind, about three meters away from my face, staring straight at me like I was a curiosity that didn't belong in its domain. I made eye contact with the bird and stood transfixed by its gaze for a few seconds. Then the bird casually drifted off on the winds, across the valley, heading due south.

It reminded me of the story of *Jonathan Livingston Seagull* by Richard Bach. Maybe, I thought, this bird was enjoying flight for flight's sake and not on a mission to find food or for any other practical reason. I glanced over the valley below and then caught sight of a second bird, very small in the distance, hovering around above our camp. The two of them then just glided off in the direction of the South Pole, further inland, over the mountains, close together.

It is interesting to note that on several occasions the early explorers also mentioned seeing skuas miles from anywhere. On Scott's 1912 trek to the South Pole they sighted a lone skua on the wing a long way inland on the polar plateau, the day before they reached their "Three Degree Depot," and both Scott and Wilson made comments about it. Frank Wild, on the Main Eastern Journey of Mawson's 1912 expedition, had no hesitation using the skua he sighted as supplementary food. "As tents were being pitched, a skua gull flew down. I snared him with a line, using dog flesh for bait and we had stewed skua for dinner. It tasted excellent."

I nicknamed that hill "Skua Ridge" and it stuck for the rest of the trip. My experience of seeing the skua had a profound effect. For me it was a sort of metaphor, which caused me to reflect on our frail human presence in Antarctica as compared to this small, delicate, warm living thing that was perfectly at home there.

Climbing alone on a rocky ledge
An island of solid rock in a sea of snow
I saw a shadow pass above me,
And jumped, startled by the strangeness
Of another life, here, in this lifeless place
A speck of warmth in time and space
The skua looked down at me
Its brown eye twinkled in the sunlight
Its glare of unconcern bathed me
As it flew away towards the Pole
Nonchalant about my plight
As it carried on its flight.

17
The Ascent of Mt. Gudmundson

Her cold white coat preserves her charm
And the cold wind's chill keeps her fresh
But still she hides her many secrets
From those who know her best.
Time will never change her
For she's constantly renewed
The source of endless mystery—
Always sought and wooed.
—a poem by Harold Brown, seaman on the *Private John R. Trowle*

Day 19, 4 December, was a clear sunny morning, 7°F. We traveled across to Mt. Gudmundson from Skua Ridge without incident, pitching camp high up on the snowy foothill of that mountain at an altitude of 1480 meters. We turned into bed early that night, readying ourselves for what we expected would be a long day. Our plan was to find a safe route up to the top of Mt. Gudmundson and spend as much time as the weather allowed working on the mountain.

It was indeed another long day. The next five pages of my notebook are filled with details of the stratigraphic section on Mt. Gudmundson, measuring each layer of rock down to every centimeter of detail to record the subtle changes in sedimentology. Fraka, our resident sedimentologist, usually did this job but because she wasn't feeling well that day I volunteered to do the job. I thought it would give me good practice at being "a real geologist" once again, taking me back to my student days as a geology undergraduate when measuring and describing rocks was an integral part of my studies. By taking such detailed measurements geologists can determine the changes in the environments in which the sediments were deposited. The layers can represent many sub-environments (called "facies") within a large environmental system. For example, our Aztec Siltstone represents a large alluvial plain system within which we could readily identify individual river channels, lake deposits, sand bars, overbank flood plain deposits, and even ancient soil horizons. Only by such careful observation of how each sedimentary layer differs, by both its composition and sedimentary structures, can a geologist determine the change from one part of the environment to another, and get an understanding of larger scale regional topography. Very detailed studies are then done by cutting thin sections of the rocks (about 30 microns* thick) and examining their mineral composition and small-scale structures under a petrological microscope.

To the trained sedimentologist, even the types of quartz present in a single layer of sandstone can reveal a wealth of information about the topography of the ancient environment in which it formed. For example, if the sediment is largely composed of quartz grains derived from the erosion of volcanic rocks, then that means the land then had volcanoes in its highland regions, and volcanoes today signify crustal rifting and volatility, and these often have ore bodies associated with them. So, it was imperative that the new section be measured and studied centimeter by centimeter, and that adequate samples of the rocks be taken at the critical points where the sedimentology varied.

However, before we could attempt any "geologizing," as Scott used to refer to it, our immediate mission for the day was to get safely to the top of Mt. Gudmundson. We could see through our binoculars that the colored layers of rock there appeared to be the fossiliferous Aztec Siltstone. Getting up Mt. Gudmundson didn't look so easy as there was a very steep ice and snow bank forming a razorback up to the rocky outcrops. These formed steep-walled cliffs of sandstone.

On our first attempt we drove the skidoos straight up the steep slope as close as possible to the rocks, but it soon became too steep for safety, so we had to level off at this point and park the vehicles. Brian

*A micron is one millionth of a meter.

suggested that we scale it in the regular way using mountain-climbing equipment, and that the two of us go up first to establish a safe route.

We donned our harnesses, roped ourselves together, fitted crampons to our boots, and each grabbed a batch of about twenty bamboo flagpoles. Brian led the way carefully up the narrow ice ridge. Each side of the sharp ridge formed a very steep icy slope. As we gradually eased up the mountain, the sides of the ridge became steeper and more distant from the rocks at the base. Towards the top of the sharp ice ridge there was approximately 300 meters of icy slope on each side of us, with large jagged rocks protruding at the bottom. Every step had to be carefully placed, so I followed in Brian's exact footsteps. In places we had to cut flat footholds with our ice picks. We placed flags along the route as we went.

On approaching the top, we encountered the vertical whitish-yellow cliffs of Beacon Heights Orthoquartzite. They appeared quite formidable at first glance, but by scouting around for a few minutes Brian soon discovered an easy climbing route around the side of the sheer outcrop that took us up above the 60-meter bluff onto the flat top of the mountain. While I stood happily on top of the cliff savoring the view, Brian quickly raced back down like a mountain goat to get the others. A short while later, Margaret, Brian, and I were standing there together. As Fraka still wasn't feeling well that day she remained back at the camp.

Step-like layers of alternating dark and light colored shaley rocks led to the summit of Mt. Gudmundson. The very top of the mountain, above these multicolored Aztec layers, was formed of a cap of about 40 meters of steeply inclined dark green glacial deposits. These are from the great Permian ice age that covered Gondwana some 290 million years ago, whose geological deposits are also found in Australia, India, and South Africa. Glacial deposits such as these are geologically very distinctive. Heavy rocks trapped in floating glacial ice drop as the ice melts, so the sediments formed below are usually a poorly sorted mixture of rocks of variable sizes and shapes in a matrix of finer grained muds and silts.

We felt the elation of every person who has ever been the first to climb any mountain, even though this was an easy one to scale and not very high at only 1950 meters! Standing on its peak gave us spectacular views of the nearby Cook Mountains and the wind-strewn polar plateau to the south.

The rest of that morning we were kept busy measuring and describing a detailed geological section through the Aztec Siltstone. Logging this section, centimeter by centimeter, took me all of that day and part of the next day, but in the course of the work several layers rich in fossil fish remains were discovered. Brian assisted me by holding the measuring staff for each locality or chipping out specimens requiring hours of patient labor. Margaret found some very interesting trace fossils that day, so she spent most of her time measuring and describing them. Later in the afternoon I made a particularly good find, a complete internal cast of a large skull of the placoderm fish *Bothriolepis.* A good deal of hammer and chisel work was required to free it from its hard sandstone repository, but it was well worth all the effort. It was the first complete skull of a fish I'd found so far on the expedition. Moreover, the internal cast of the skull shows some of the most diagnostic features for the different species of *Bothriolepis*, so I knew at once that the specimen would be useful for identifying its exact species. We kept working on top of the mountain until about 8:00 P.M., and we all collected loads of great specimens. Our backpacks were bulging and heavy by this time, so we carefully headed down along the flagged route back to the camp, arriving just after 9:00 P.M.

Fraka had spent the day resting in the tent and was feeling much better by the time we returned to camp. She was excited to hear about our day's work and agreed to join us up the top of the mountain the following day.

I awoke to a "warm," hazy morning with the temperature at 18°F, and light winds blowing at six knots. Overhead hung a cloudy grey sky with the possible threat of snowfall. After breakfast all of us ascended the mountain and by working as a team we finished logging the section by mid-morning, now aided by Fraka's expert sedimentological eye. I then went off to collect fossils from each of the layers we had identified the day before. Many excellent specimens turned up that day, including more *Bothriolepis.* Some layers were quite rich in small fish remains, like the interesting double-pronged sharks' teeth and the fin-spines of acanthodian fishes.

Margaret was overjoyed to find a very rare "resting trace" of the creature that probably made the *Beaconites* traces, as it was sitting on top of one of these distinctive burrows. Such fossil burrows are known from many sites throughout the world, but never before had anyone found any actual fossil evidence as to what kind of animal might have made the burrows, so this was an exciting discovery. Having no materials on hand to make a cast of the trace, Margaret attempted to make an "ice cast" using her personal supply of orange juice drink. She first covered the trace fossil with a layer of thin wrapping plastic, then poured her thermos of orange drink over the top. She returned later that afternoon when it had frozen solid. Unfortunately, it broke apart when she tried to lift off the ice cast, so she had to content herself with just photographs and detailed sketches of the important discovery. We discussed the trace at length in the field, and agreed that it most closely resembled the body plan of a flat lobster-like crustacean, maybe a primitive relative of today's delicious Moreton Bay bugs (a small lobster-like creature indigenous to Queensland, Autralia).

The next morning felt almost hot when the temperature soared to 21°F, with only a faint wisp of a breeze. In the sky some lenticular high-level clouds were suspended over the dark brooding Warren Range, about 60 kilometers away from our camp on the other side of the Mulock Glacier. It was time to make tracks, so we packed up camp and headed back to our depot at Seay Peak in the northern Finger Ranges, arriving in the early afternoon after a pleasant and uneventful sledge journey.

Our next task was to prepare our gear for the helicopter lift across the Mulock Glacier due to take place the next day. We had to make sure that all our specimens were carefully wrapped, labeled, and packed in wooden boxes for transport back to the base. Then every piece of equipment, including our food supply boxes, had to be accurately weighed so that we could tally the full weight of each underslung load. We did this using a set of small hand scales. The skidoos were to be underslung by the helicopters, so they had to be stripped down. We removed their fiberglass fairings, secured the steering mechanisms and emptied any remaining fuel out of their tanks. The bare sledges would be tied to the sides of the landing gear of the choppers.

The helos would also take "retro"—the term used for anything to be taken back to Scott Base. This comprised our samples collected so far, our rubbish bags and our ominous bag of frozen human waste. There's an interesting story about this subject told to me by Brian how a few years back the helicopter pilots used to keep their cabins nice and warm by turning the heaters up high. One time a helicopter came to transport a field party and take back their retro, which included a large plastic bag of frozen waste. Someone carelessly threw the bag in the helo behind the pilot's seat, and the crew commenced their long flight back to base with the heater blaring away inside their cabin. After a while they could smell something really awful and soon had a lot of unsavory brown slop swishing around their feet, as the bag had ripped apart and the waste material had melted from the heat. From then on, VXE-6 Huey pilots always arrived at such missions wearing thick warm clothing and refrained from heating up the cabins of their helicopters on trips back to base. The bags of refuse were stored at the base frozen outside near the sheds until we returned, when we would dispose of them by incineration.

That night we discussed the crossing of the Mulock Glacier. Not only did we hope to receive news from home when the choppers arrived, but we would be getting re-supplied with some fresh foodstuffs from Scott Base. It also signified the end of the first half of our field trip, our successful journey through the Cook Mountains from the Darwin Glacier at 80° south to the Mulock Glacier at 79° south.

After our airlift over the Mulock Glacier we would be able to start exploring the many mountain ranges and isolated outliers fringing the vast white plain of the Skelton Névé, where earlier VUWAE parties had collected superb fish fossils during the 1970-71 season. We were full of hope that new fossil localities would be found in the region during our explorations.

A new journey was about to begin.

18
Over the Mulock Glacier

Out of whose womb came the ice?
And the hoary frost of Heaven, who hath gendered it?
The waters are hid as with a stone,
And the face of the deep is frozen.
—The Book of Job, 38:29

After his ship, *Endurance*, was crushed in the pack ice in Antarctica in 1915, Ernest Shackleton tore this page out of his bible and kept it. Perhaps he felt the biblical reference to the ice was something worth remembering, something to keep him going through the long, arduous struggle for survival upon which he and his men were about to embark. It worked. They lived. The last line, though, is a paradox that reflects the opposite condition to Antarctic seas, in which the surface is often frozen but the deep is liquid. The "face of the deep" mentioned here actually brought to my mind the image of the Mulock Glacier, whose great depths, maybe a kilometer or more to bedrock, would indeed be frozen and solidly compacted.

Two Hueys, a bright orange one from the US VXE-6 Squadron and a dark camouflage green one from the Royal New Zealand Air Force, suddenly arrived on Saturday, 7 December at 10:30 A.M., blasting snow and noise all around our dismantled camp. The tedious work of the previous day preparing the sledges, skidoos, and samples for return to Scott Base was now behind us. It only took two trips using both helos to move all our gear and us over the glacier.

The Mulock Glacier is about 50 kilometers wide at this point. It has the most enormous crevasses I'd ever seen in its center, formed as the wide, flowing river of ice bends around the Warren Range to the north and the Finger Ranges to the south, and then expands outwards as it nears the sea. This expansion of the compressed river of ice causes crevasse fields to develop. From the air the crevasses appear to be between five and ten meters wide in regularly ordered sets, oriented perpendicular to the flow direction of the glacier. It would be absolutely impossible to try and cross it on the ground without imminent disaster.

After the helicopters had dropped our gear on the other side and departed, we had to first reassemble all the equipment and then repack the sledges. As the site where we were dropped was in the middle of the Deception Glacier, on hard blue ice with no shelter from the winds, we decided that we should immediately move on towards the first of our new destinations, Mt. Ritchie, about ten kilometers southwards down the glacier. The plan was to come back this way after we finished working at Mt. Ritchie, so we could leave excess equipment and supplies behind here and travel light, taking only some food, fuel, and tents loaded onto one sledge each. I enjoyed traveling this way as it was far easier to handle one sledge rather than two, which were sometimes quite difficult to maneuver when using the very long ropes between each sledge required for traveling in crevassed areas.

Although it was only a short distance to Mt. Ritchie, it soon became a rather gnarly journey. At one stage our sledges went over a small crevasse, the runners gliding silently over a snow-covered gap that opened up underneath it. Brian then stopped to fetch the two-meter long crevasse probe. He walked slowly ahead, feeling his way through the small crevasse field, eventually leading us onto safe ground again.

Later that evening Brian and I, who were then leading the way, found ourselves going down a very steep ice slope. Brian was driving the Skidoo, while I was on the back of the sledge. I became aware that we were steadily moving along faster and faster as the slope increased, making it difficult to steer and brake the sledge. I suddenly realized with a sinking heart that the sledge was actually headed down a steep ice slope and the brake wasn't having any effect on slowing it down. My sledge was suddenly racing out of control down towards piles of jagged rocks at the bottom of the slope. Brian saw this and immediately powered up the Skidoo to go faster, keeping pace as I accelerated, not able to slow it down nor turn it away as we were on dangerous ground—solid ice. I kept smashing down hard on the ice brake but at that speed the metal teeth of the wooden brake only bounced off the jagged blue ice and jarred my foot. It seemed hopeless. If the rope suddenly tightened the sledge would have flipped over and hit the blue ice with its protruding stones at a very high speed, probably smashing it to pieces. Brian skillfully kept pace next to me all the way; the minutes seemed like forever as we raced at deadly speed towards the rocks ahead. Finally, the slope eased off and I was able to steer the sledge sideways, away from the rocks in front until it came to a grinding halt at the base of some rocky scree at the foot of the mountain.

It was about 8:00 P.M. and I was quite shaken up by the incident, mainly because for the first time on the sledging journey I had no power to control the sledge. Furthermore, we couldn't see how steep the slope ahead was gradually becoming. For all we knew it could well have carried on getting steeper, ending at an icefall or in a crevasse field. Luckily for us this incident had caused no harm or any damage to the equipment.

We were all exhausted by this time, and I recall there was some minor squabbling, mainly due to our tiredness, over where we should pitch our camp. It was the first time so far on the trip that we argued amongst ourselves, mainly because some of us were so tired that we just wanted to pitch camp anywhere, but another member of the group insisted that we take the time to find a good camp site. Eventually we all agreed on a spot and quickly put up the tents in a strengthening cold breeze. Margaret was soon busy making a meal for us. Not long after enjoying our food we were all ready to sleep. However, one thing that happened that day kept us in good spirits and made us laugh heartily over dinner whenever we thought of it.

For the last few weeks I had been getting up out of my sleeping bag every morning in the freezing cold, putting on a few layers of clothing and boots, then rushing outside for my morning leak. I eventually noticed that my tent mate, Brian, never did this. Instead he always slept in and had a knowing, sly grin on his face over breakfast each morning. Finally I questioned him one day as to what his secret was.

"Do you have an iron bladder or something?" I asked him.

"No," he replied calmly, taking the stumpy screw-top plastic bottle out and handing it to me.

"What is it?" I asked naively.

"It's a piss bottle," he replied with a grin, his Kiwi accent emphasizing the adjective.

"What do you do with it?" I asked even more stupidly, already suspecting the obvious answer.

"You piss into it, screw on the lid, then tip it out into the snow between the flaps of the tent. Saves you having to get up every morning for a piss."

"Should have told me about it before!" I replied.

He grinned from ear to ear. I could see now that it was his little joke, and he seemed to enjoy watching me scramble out into the cold every morning. Maybe, I thought to myself, it was a secret jealously guarded by the hardened, long-experienced deep fielders, one they only reluctantly passed on to new boys like myself.

Brian said that we could ask for one on the radio schedule next morning. As we were going to get new supplies from Scott Base on the day we crossed the Mulock Glacier, this would be the time I would finally get my piss bottle. I couldn't wait!

Next morning we were going to request the bottle. I grabbed the radio mike and heard a woman operating the radio from Scott Base. I felt a little embarrassed about asking for a "piss bottle," so wasn't sure how to proceed. I was giggling over it for some time, thinking of what to say, when Brian grabbed the mike and calmly asked the female operator on the other end if, on the forthcoming re-supply, we could add one more item to our list of essential new equipment?

"John needs a drinking bottle," he said smiling at me. "It must be a wide-mouthed one with a screw top lid, so we can pack it full of snow."

I was rolling about the tent laughing at this, but Brian just smirked wisely, clearly pleased with his disguised way of ordering the item.

We made a regular joke about this during the several days leading up to the re-supply, making a special point to highlight "John's wide-mouthed drinking bottle" on each checklist.

Finally the big day came. We learned that Dave Geddes, then head of the New Zealand Antarctic Program, was coming out in the helicopter to visit us. It was a great honor and we were looking forward to seeing him and passing on the good news about how well the work was going.

I'll never forget that glorious moment. We hadn't seen another soul for over a month, and as soon as the chopper came in and landed near us, Dave Geddes leaped out like a crack SAS commander, head tucked low under the whirling rotor blades, as he sprinted over to us, smiling from ear to ear.

"Here, John," he said, enthusiastically shaking my hand while thrusting the plastic bottle into my other hand, "here's your water bottle."

We all fell about the place laughing hysterically. He had no idea why we all cracked up. He must have thought we were all going a bit snow-crazy.

The bottle had a little plastic drinking straw coming out the top.

19

Mt. Ritchie and Deception Glacier

Distant mountains floated in the sky as enchanted cities, and often the whole white world would dissolve into a gold, silver and scarlet land of Dansanian dreams and adventurous expectancy under the magic of the low midnight sun.
—H.P. Lovecraft

The black, brooding Warren Ranges draped in a light snow mantle, its peaks hung with stagnant fluffy clouds, is truly an awe-inspiring sight to behold. Add the hazy midnight light of a clear summer Antarctic night, and one can empathize with Lovecraft's timeless words.

It was lightly snowing when we woke up on the morning of Sunday, 8 December, the temperature at a comfortable 18°F. After chatting with the others, who wanted to stay inside and rest for the morning, I decided to head up Mt. Ritchie alone to search for fossils. Mt. Ritchie was one of the most important fossil sites in the whole region, the thickest outcropping exposure of Aztec Siltstone that had been discovered so far.

I packed a good variety of snack foods and a thermos of hot orange drink in my large backpack, then bundled up in my thickest gear in case the weather turned foul. At 9:20 A.M. I left the camp in a fresh 20-knot wind and headed off, slowly pacing up the treacherous rocky scree slopes. I noted the first *in situ* fish fossil bed about two hours later, when I was standing some 30-40 meters down from the top of the exposed section. This was a totally unexpected find as the previous VUWAE expeditions had only found fish fossils at the very top of the mountain. I wrote down the locality with sketches of the cliffs above and below me to try and accurately determine the exact position of the new finds. After collecting a few samples I decided to make more of this site on the way down, just in case better material was waiting ahead of me. I had learned by now to take the chance and push on in search of new discoveries if the weather was on my side. At that time I was awfully suspicious of just whose side the weather was on, but took the risk anyhow.

It was tiring work walking up the rubbly ice-cemented scree because it was loose and moved easily underfoot with each step, but I was excited at seeing so many fragments of fish fossils almost everywhere around me. It took about three hours to reach the top of Mt. Ritchie. The summit consists of a near-perfect pyramid of black volcanic rock sitting on greenish Permian glacial deposits.

I had also been assigned another mission that day: to find a long-lost pen. When Alex Ritchie was there in 1971 with VUWAE 15 he lost his silver Parker pen, a special gift from his mother, so he had asked me if I would have a good look around for it on the top of the mountain. After several trips to the peak all I found were some scraps of newspaper, dated 1970, which they must have used to wrap their fossils; but alas, no silver pen. Later, I learned from Alex that I had my localities mixed up as he had lost his pen on the mountain range near Alligator Peak. The buried treasure of the Ritchie clan resides there to this day.

But the treasure I did find was in the form of a beautifully preserved large set of fossil fish jaws belonging to an acanthodian. Acanthodians were spiny, shark-like fishes that died out at the end of the Paleozoic Era some 250 million years ago. These jaws are a rare find and my preliminary study of them indicates that they could belong to a new undescribed species. The upper and lower jaws were preserved together in natural articulation, and I suspect that probably the whole fish, maybe a meter or so long, was once buried there 380 million years ago, possibly in a shallow lake environment. Since being exposed through the earth's upheavals on the top of a mountain in Antarctica, the rest of the fish's body had now unfortunately been eroded away. The harsh winter bombardments from ice and sand particles in fierce storms, and from the freezing and thawing action of ice, quickly break up layers of fine-grained rock. Many of the surfaces of the rocks there had a polished appearance that testified to the extreme power of nature that reigns during those winter blizzard months, year after year, for countless thousands of years.

Despite the wonderful discovery, the day did not go as smoothly as I'd expected. Many times I slipped over on the steep icy slopes, losing my footing and crashing down on the rubble. At one time I came to sheer vertical cliffs and could go no further, so had to backtrack a fair way to try another approach. By the time I was into the richly fossiliferous layers near the top of the mountain, I rapidly filled up my backpack with specimens, which soon became very heavy. It was like being a little kid in a candy store. There were so many specimens I wanted to collect but I only had limited time and enough space to bring back the most important finds.

Somewhere near the top of the mountain I found myself on a steep ice slope. I tried to get to some prominent beds of sandstone separated by a thick layer of compacted icy snow. Eventually, by digging foot holes in the ice with my ice pick, I was slowly able to crawl up to reach the summit where the sandstone ledge jutted out. I spent some time following this ledge until it petered out, then found I was surrounded by steep snowy slopes below me, with rocky ledges maybe ten meters further down. Rather than go all the way back, I could see some interesting outcrops further below, so began carefully stepping onto the white slope. Instead of finding soft snow, which I expected to sink into, it was mostly solid ice with a fine snow layer on top. I immediately lost my footing, slipped and, burdened with the heavy backpack full of rocks, found myself sliding down the slope, rapidly accelerating towards the rocks. I braced myself, then crashed heavily into the rocky ledge. Dazed and sore, I got up, straightened myself out, ate a little chocolate, as one always does in times of stress in Antarctica, and kept moving on. There was a continuous dull aching from my legs and side that remained for the rest of the day.

It hadn't stopped snowing all day and the weather appeared to be rather unpredictable. At this point, late in the afternoon, I decided to call it a day and head back to camp. I arrived safely back two hours later, a bit tired, my legs aching from carrying a really full pack loaded with fossils and my body still sore from the fall. Nonetheless, I was pleased with the day's discoveries. It had been a most exhilarating day to say the least.

That night we ate very well, fuelled by "freshies," the fresh foods dropped in from Scott Base at the re-supply, which incidentally included a reasonably decent bottle of Australian red wine, a Saltram's Cabernet Merlot.

The weather had grown worse by the next morning. Icy winds gusted at between 20 and 40 knots. Brian, Margaret, and Fraka decided to stay at camp again, so I packed my gear to head up the mountain alone once more. About one-third of the way up I made an unexpected discovery of another layer of fish fossils, surprisingly low in the section. Amongst the specimens I collected there were fragmentary plates of phyllolepid placoderms, an interesting discovery that hinted at the group being further down in the section rather than restricted to the top few meters, as I had previously read. I also collected some superb acanthodian fin-spines and various isolated bony plates of the placoderm *Bothriolepis*, including some very large plates of the gigantic species *Bothriolepis mawsoni*, named in honor of Sir Douglas Mawson by my colleague Gavin Young. *Bothriolepis mawsoni* was the largest of all the southern species of this genus. Its huge bony plates suggested this fish was almost a meter long from snout to tip of tail.

In the early afternoon I climbed to the peak of Mt. Ritchie, and standing there was briefly able to enjoy the splendid view out over the Skelton Névé. The winds were now howling around me, dangerously strong, so I had my lunch tucked tightly in a wind shelter, a hollowed out cavern at the base of the Permian Sandstones. I made myself a dehydrated lamb meal by adding hot water from my thermos to a packet of dried food, and although it was still a bit chewy it went down a treat.

I spent the rest of the afternoon collecting more specimens from the top layers of Mount Ritchie before returning to the newly discovered sites in the lower layers that I had found on the previous day. Here, 120 meters from the base of the section, was the lowest layer so far discovered at Mt. Ritchie containing fish fossils, so I suspected the faunal assemblage should be quite different from the better known fish faunas at the top. This turned out to be true as no sharks' teeth had been found from the top layers, yet these were quite common in the lower sites.

That night I returned once more triumphantly loaded with a full pack brimming with fossils and we were all in high spirits as we could feast once more on the fresh food supplies. We ate fish served with a salad and baked potatoes, followed by a dessert of real strawberries in mock cream (made up from powdered milk). I slept deeply that night in anticipation of another day's collecting.

We awoke to an average kind of day, about 14°F, an overcast sky and moderate winds. All of us were all keen to go exploring for new fossil sites, so our immediate plan was to first make a quick reconnaissance trip over to the Boomerang Range, about eight kilometers away from our camp towards the southern end of the range. On arrival we quickly found that the only outcrops of Aztec Siltstone were exposed facing out to the Skelton Névé. There was no easy access to get to these layers from the Deception Glacier because of a very steep ice cliff on the other side.

Huge gusting winds, roaring up off the cliff face of the Boomerang Range, blasted billows of fine powdery snow high into the air like a tormented geyser. In the lee of the ridge, though, it was quite still. Brian walked over to the blasting wall of snow and looked over the edge of the scarp. He braced himself and could only peer momentarily into the strong winds. We decided then to work elsewhere, so we rode over to the other side of the glacier to some rocky bluffs just south of Mt. Ritchie. One of these small cliff sections, informally named Mt. Kohn by VUWAE 15, had yielded some interesting fish fossils, so we decided to search there and measure a section through its exposure of Aztec Siltstone. The steep ice slopes on Mt. Kohn, combined with the winds increasing in strength, required us to use instep crampons over our boots to get up the scree slopes.

I scrambled up most of the 130 meters of outcropping Aztec Siltstone here and found some good specimens, including a fossilized impression of a distinctively ornamented acanthodian cheekbone belonging to the genus *Culmacanthus*. *Culmacanthus* was the first new genus of fossil fish I had ever described, back in 1983. It was based on some very well-preserved specimens from Mt. Howitt, in the central mountainous region of Victoria, and I named it after the Aboriginal word *culma*, meaning "a spiny fish" because of its enormous fin-spines. Gavin Young first discovered *Culmacanthus* plates in Antarctica and placed these in a new species, *Culmacanthus antarctica*, in 1989, further emphasizing the close links between the Australian and Antarctic fossil fish faunas. To date, the genus has only been recorded from these two continents.

Mt. Kohn has an interesting story attached to its name. Barry Kohn was a geologist who was deputy leader of the 1970-71 VUWAE 15 expedition on which Alex Ritchie and Gavin Young collected fish fossils. Anyhow, while working on this treacherous slope, which is very steep and icy as I mentioned, Barry was unexpectedly blown off a ledge near the top of the section by strong winds. He fell about 30 meters, bouncing down the jagged scree slope. He was badly cut and broke his collarbone, so had to be airlifted out by helicopter. He recovered back in the hospital at McMurdo Base and later rejoined the expedition. Since that day the mountain was nicknamed by the VUWAE team in his honor as "Mt. Kohn-descending."

We tried to get to the western side of Mt. Ritchie later that afternoon but there were too many crevasses, so we headed back to camp. The following day was fine and warm, at 18°F. After packing up camp we sledged the ten kilometers north from Mt. Ritchie to our depot at the base of the Warren Range, taking a wide course around the small crevasse field that we had stumbled into on our way down. The retro to go back to base was left alongside some spare drums of helo fuel, making a neat pile of boxes and waste bags to be picked up by the choppers at some later date.

With the sledges repacked we set off up the Deception Glacier around 4:30 P.M. The south end of the Warren Ranges didn't look too promising for us because snow covered much of the rocks, so we decided to leave this section and move on further north. It soon became difficult towing the now heavily laden sledges up the glacier. One sledge carried all the fuel drums, seven in all, each weighing 64.5 kg; the other sledges carried over eleven food boxes plus the kitchen boxes, tents, and our personal gear. Despite having to relay sledges up the slope at the head of the Deception Glacier where the slope was too steep, we had an easy journey back and arrived at a good camp site at around 8:00 P.M. not far from Moody Peak.

It was indeed a very scenic place to spend the night. Dark dolerite mountains of the Warren Range had a light snow cover that accentuated the flow banding in the rock. This is a feature of volcanic rocks formed by separate episodes of molten rock cooling in stages, a result of convection currents within the molten rock mass. The strongly contrasted white layers of fresh snow on the towering jet black rocky crags was an extraordinary site when coupled with the sun shining from a hazy blue sky punctuated with occasional fluffy cumulous clouds.

After dinner that night we stumbled upon an amazing discovery: how to make the drink Irish Cream by mixing very cheap Scotch whisky with a quantity of thickly reconstituted powdered milk and sugar. We dubbed this drink "Deception Irish Cream" mainly because we were camped on the Deception Glacier, and also because we couldn't tell the difference between our concoction and a certain well-known brand of the same drink. . . .

This recipe is included in Appendix 2.

20
Onwards to Escalade Peak

There was something hauntingly Roerich-like about this whole unearthly continent of mountainous mystery. I had felt it in October when we first caught sight of Victoria Land, and I felt it afresh now. I felt too, another wave of uneasy consciousness of Archaean mythical resemblance; of how disturbingly this lethal realm corresponded to the evilly famed plateau of Leng in the primal writings.

—H.P. Lovecraft

This passage from Lovecraft reminded me of the most overwhelmingly beautiful scene I have ever seen, which we encountered quite by accident on this next part of the journey. While standing atop of a ridge near Escalade Peak, I turned around and saw an amazing vista of the Transantarctic Mountains framed by deep blue sky with horizontal layers of white cloud at two separate levels around their mystical peaks. The sun shone down in widely radiating rays, giving the whole vision an aura of transcendental power, like a scene painted specifically for inspiring religious zeal.

It was a busy day on Thursday, 12 December, as we moved on from Moody Peak down to the vast open plateau of the Skelton Névé, leaving two of our sledges with supplies at a storage depot at the head of the Deception Glacier. The first part of the journey was fraught with terrible sastrugi: they were rough, jagged, and very high. It was slow going. We had five overturned sledge accidents—every one of our sledges suffered an overturn or two, but luckily no significant damage was incurred.

After lunch we moved swiftly to the Swartz Nunataks, east of Escalade Peak. It was relatively warm, 12°F at noon. We then swapped teams, Fraka and I taking one sledge train, Margaret and Brian the other. Fraka drove the Skidoo all morning; then I drove from the depot in the late afternoon. After we had made it through the rough sastrugi we entered a wide-open flat snow plain, almost perfect conditions for sledge travel. We were able to scoot along at 20 kilometers per hour from this point on almost to the end of the day. It was an uplifting, exciting feeling to drive the Skidoo pulling only one sledge over a perfectly flat smooth surface. We reached our destination by 8:45 P.M. Although we had completed 57 kilometers of sledging that day, the first ten took us four hours and much hard work to upright all the overturned sledges. We had now covered more than 400 kilometers since the beginning of our trip.

I cooked us a quick dinner that night of dehydrated beef curry with rice, served with soup and bread. Using the dehydrated meals was always preferable after our long heavy days of traveling when we were all dog-tired and needing a quick feed. Dinner was ready by 10:30 P.M. and finished at close to midnight.

The next day was black Friday, 13 December, although I must admit it was the whitest black Friday I'd ever experienced. It had been exactly four weeks to the day since we were put in the field. I'm not superstitious in any way but, for some reason, I thought to myself that I would not take any additional risks that day. It started off as a warm sunny morning with a temperature of 10°F, and light 7-knot winds. We were camped near an outcrop of sandstone at the Swartz Nunataks that Margaret wanted to examine. It was also scheduled that day for helo NZ1 to pick up the fuel and retro from the Deception Glacier and call over to us to drop off some extra lashing ropes, mail, and possibly other goodies.

We passed a few hours searching the low hill of greenish-grey sandstone outcrops near the camp. Margaret concluded that that was probably the Junction Spur Sandstone, a rock unit that we had studied at the base of the succession near Gorgon's Head and at Skua Ridge, so we knew it was unlikely to contain any fish fossils. Escalade Peak, a prominent range over in the distance, looked far more interesting from where we were camped as we could see that it sported an extensive sedimentary sequence of lighter colored rocks.

Unfortunately, the helo couldn't come out that day as we heard on the evening radio schedule that the weather had turned foul shortly after it had set out, so they had decided to turn back while they could. Apart from this one element of bad luck, it was a rather uneventful black Friday, just the way we liked it in Antarctica. I thought to myself that our careful planning and faith in science had once more dispelled superstition.

The following morning was gorgeous, with the temperature at 16°F at 8:00 A.M., and it remained warm, sunny weather all day. In good humor we packed up camp and moved swiftly on to Escalade Peak, arriving there uneventfully at 3:00 P.M. After setting up the new camp we had a quick Skidoo run up to the saddle between Escalade and Tate Peaks to check out a possible plan of crossing down through the saddle to head on to the Boomerang Range later. The area below the saddle turned out to be riddled with large crevasses, so our only option for getting to our next site was to go round the long way, backtracking on our route.

Following dinner that night Margaret surprised us by producing a homemade cake which went down a treat. That night we invented yet another new cocktail: "Mock Kar-lua" (the recipe for which is also in Appendix 2). We then played a game that I introduced to the group where we would have readings from my little pocketbook of famous quotations, and everyone had to guess who was the originator of the quotation. The game became a popular recreation for us at nights for the rest of the trip.

I wrote a short health report in my notebook that day as it was our thirtieth day in the field:

> So far, so good. Skin a bit scaly on legs and arms, cracked around fingernails; lower lip is blistered and split, almost healed now (has been like this since Nov. 28th). Teeth, mouth fine (regular brushing); nose a bit burnt, but O.K. Strength good, stamina improving; hands get cramped after driving the Skidoo for a day. All else fine.

On Sunday we worked up on the small outcrops of Hatherton Sandstone in the saddle between Tate Peak and Escalade Peak. There were lots of excellent trace fossils exposed here. I stumbled upon one of the most amazing finds of the whole trip that day. On racing up to the outcrop I was the first there to wander around and examine the large slabs of Hatherton Sandstone. Suddenly I laid eyes on a humungous set of animal tracks, somewhat resembling two motorcycle tire tracks in parallel. They were a set of giant arthropod tracks beautifully preserved on a large flat slab of yellowish sandstone. The track way measured 84 centimeters wide with individual appendage traces up to 19 centimeters across. This sure was the mother of all arthropods. A few years earlier Margaret discovered similar large tracks from Perseus Peak near Gorgon's Head. She was excited by the new find and spent several hours studying it, even managing to chisel out a part of the opposite side of the track way (the positive mould), which she found nearby. She believes they were made by giant sea scorpions (called eurypterids) that grew to about two meters long from head to tail! These beasts lived largely in the sea and marginal river basins, where they hunted prey with their formidable crab-like claws.

The best-preserved tracks on this new slab showed individual segmented claw marks where the three legs on each side were overstepping each other. The 19-centimeter-wide trace had three clearly visible sets of digit marks. These monstrous sea scorpions occasionally emerged out of the water to venture onto dry land. Evidence for this is seen in well-preserved tracks in the Tumblagooda Sandstone exposed in the scenic gorges near Kalbarri, Western Australia. Here the tracks sometimes show fossilized "drips" of wet sand around them, indicating that the animal was completely out of water when it walked. In some places at Kalbarri there are also "slide marks" where the beasts were sliding down the muddy banks on their bellies into the water.

The age of the Hatherton Sandstone in Antarctica is still controversial. Some scientists, like Dr. Nigel Trewin of Aberdeen University, think it could be Early Devonian (about 400 million years old) because there are similar trace fossils to be seen in the Old Red Sandstone of Scotland. Yet recent findings from Australia on the age of the Tumblagooda Sandstone suggest that if the Hatherton Sandstone contains the same trace fossils as the Western Australian site, then it could be much, much older. The rich trace fossil assemblages of the Tumblagooda Sandstone have much in common with the Hatherton Sandstone. Both have *Heimdallia*, an intensely burrowed trace in the rock (looking like spaghetti rock), plus similar forms of arthropod tracks called *Diplichnites* and *Beaconites* burrows.

Quite recently it has been suggested that the Tumblagooda Sandstone could be as old as Late Ordovician age (about 440 million years ago), based on conodont microfossils. These are miniature jaw-like structures that come from a free-swimming worm-like animal, actually more closely related to fishes than to any of the other invertebrates. These fossils occur above the Tumblagooda Sandstone in limestone. If this is the case, then this is clearly the earliest evidence in the world of arthropods leaving the sea and walking on dry land. The footprints of the giant sea scorpion at Escalade Peak show a remarkable degree of topographic relief, showing the pushed up mounds of sand around each "step," indicating an animal of huge weight sunk down into the soft sands as it lumbered along.

After lunch that day we climbed up the other side of the saddle to examine the base of the Beacon Heights Orthoquartzite, which formed almost vertical bluffs on Escalade Peak. Here we found a few trace fossils, but noticed how it graded into coarser sedimentary layers towards the top. In places the usually fine-grained Hatherton Sandstone appeared to have pebbles up to four centimeters wide scattered throughout, suggesting that it was a slightly shallower environment of deposition, perhaps heralding the transition from marine to freshwater habitats in the ancient sedimentary succession.

The amazing thing about being a geologist is how you can read the changes in the rocks, and then see how that environmental change reflects the changes in animal or plant communities of the time. It's just like being a fortuneteller in reverse. In this case it showed a clear decline in abundance and diversity of life forms, as indicated by their diminishing trace fossils. The steep bluffs of whitish-yellow Beacon Heights rocks were largely devoid of any fossils except near the top, where they sometimes contained rare fossil plants and scant fish remains. This indicated a clear transition to terrestrial conditions. In simple terms, we had left the sea and were now standing upon Devonian land. Had we been there 380 million years ago it would have just taken us a few minutes to throw in a baited line and do some serious fishing. Instead, we reached for our geology hammers and had to be content with merely scratching away in the crumbly sandstone layers in the hope of landing a few fragmentary fossils. Unfortunately we found nothing of interest at that site, as we were still well below the base of the Aztec layers.

That night we all turned into bed early with expectations of another shift of venue the next day.

21

"I'm Dreaming of a White Christmas"

We were up at 11pm, but so much time was absorbed in making a special stew for Christmas from some of the bones that it was not until 2:30 am that we got under way. To make the spread more exceptional I produced two scraps of biscuit that I had saved up, stowed away in my spare kit bag, as relic of the good days before the accident. It was certainly a cheerless Christmas; I remember we wished each other happier anniversaries in the future, drinking the toast in dog soup.

—Douglas Mawson

What a contrast this passage shows to the lavish Christmas dinner I enjoyed on Scott Base in 1989. Our Christmas dinner in 1991 would be far more frugal due to unforeseen delays from bad weather, but nowhere near as bad as Mawson's Christmas in 1912!

We moved from Escalade Peak to the southern end of the Boomerang Range on 16 December, traveling more or less in a giant U-shape to avoid crevasses that radiated out from the base of the mountain peaks. It was another long day of traveling that involved crossing two crevasse fields, both unmarked on our maps. Although we encountered some large crevasses, Brian's skilful guidance kept us from penetrating further into these dangerous areas by steering us out more towards the center of the Skelton Névé.

Large scalloped sastrugi hindered our progress between Escalade Peak and the Boomerang Range, causing one sledge to topple over twice. One of these overturns threw the first aid box off the sledge. It broke open, scattering various bottles and packets of tablets onto the snow. This also caused some minor damage to the handlebars of the sledge. Brian drilled out the broken bolthole by hand, inserted a larger bolt and made a new bracket for the wire frame. These repairs held us up for about an hour. At 7:00 P.M. we arrived at a spot about four kilometers out from the face of the Boomerang Range, not far from where we had seen the huge spouting fans of snow blasting up off the cliff faces about a week earlier.

The next day was to be the first in a long string of frustrating days involving bad weather, intermittent bad weather, and just waiting around with nothing to do while the hazy near whiteout conditions alleviated. We were now camped at an elevation of 1,300 meters. The temperature that morning was 14°F and winds blew gently from the south at around 5 to 10 knots. We were expecting helicopter NZ01 to fly in later that morning to drop us more food and mail plus some new lashing ropes, and to possibly to pick up our mail for home, if the weather was suitable for a helicopter landing.

Outside it soon became rather sticky, a mixture of dense, powdery snow, and light wind creating poor visibility with virtually no ground definition. A thick blanket of snow covered the mountains that we wanted to search, so any ideas of working there were dismissed until conditions improved. The day was spent mostly reading. I did a pencil sketch of the mountains that were visible from our camp in my notebook, also noting that it had become much colder that day with an increasing wind-chill factor.

The helicopter came that afternoon, hovering just above the ground to drop off our supplies and mail, as the pilot didn't want to risk landing due to poor ground definition, so unfortunately we weren't able to send any of our letters back home.

I received a tape of music and talk from my family that day, plus a letter from Gary Morgan, then the Curator of Crustaceans at the Western Australian Museum. His letter relayed the necessary paperwork I needed to bring home a giant marine isopod specimen I had been given for the museum collections.

Margaret loaned me her Walkman to listen to my tape from home. Eagerly I curled up in a corner by myself and listened to it, relishing every word of my wife and children's voices. They had selected a series of my favorite songs as well as humorous pieces. There were also songs sung by my own children, Sarah, Peter, and Madeleine. It brought home to me how much I missed them all and, in all honesty, more than a few tears flowed during that marvelous hour, especially when my youngest daughter, Maddy, then only three years old, told me that she loved me and missed me and hoped I'd be home soon. It was the best possible gift one could ever get. It had some sort of profound effect upon me as later that afternoon I had a wash, including my hair, and changed my underwear. This was the first time I'd done this since we set out on this trip 32 days before!

Having a full body wash involved first firing up both primus stoves to get the ambient air temperature inside the tent as high as possible. Then, after heating up some water in the camp oven, I stripped off so that I could quickly wash myself with soap and a warm sponge. After toweling myself dry I squatted down over the camp oven and scooped hot water over my head and applied some shampoo, and later rinsed it off back into the camp oven. I was able to use the leftover warm soapy water to wash my underclothes and socks in. The washed garments are rinsed in a little fresh water and hung outside with pegs over the tent ropes to freeze solid. Drying winds slowly ablate away the ice from the material, leaving them soft and dry, although still very cold, by the next day.

Continuing poor weather conditions greeted us the next morning. Outside visibility was hazy due to a light snowfall. I felt compelled to examine the outcrops of Aztec Siltstone exposed in the southern end of the Boomerang Range that I could see through the binoculars, although access to the outcrops appeared to be difficult from where we were camped.

I grabbed a shovel and began making a cleared roadway down to the blue ice by leveling out the intermittent sastrugi, some of which were about half a meter high and very irregularly shaped. A line of flags on bamboo poles was then placed to mark the route in case visibility became worse. I tested the new route by slowly driving the Skidoo over it. As it was fine I went back to camp and told the others I was going over for a small reconnaissance trip to the mountains by myself. Armed with a few handfuls of food and a thermos of drink in a backpack, I slowly headed out across the blue glacier ice on the Skidoo. Periodically I'd stop to place flagpoles in cracks between the ice, or sometimes propped them up with small cairns of rocks. Such precautions are vitally important should the weather suddenly turn foul as visibility could be reduced to almost nothing. In this case the flags and my compass bearings would be my only recourse to find my way back to the camp.

It was only four kilometers to the other side of the glacier. Rocky moraine formed a scree slope leading to the base of the mountain range. After searching for an easy way up to the outcrops, it became apparent that there was no safe access route onto the mountain. The outcrops of the Aztec Siltstone were very high up from where I stood, possibly another hundred meters or so. The base of the mountain was formed of almost sheer vertical rock walls, a characteristic feature of the Beacon Heights Orthoquartzite that nearly always underlay the beckoning Aztec layers.

I had a quick search in the loose rocks at the base of the range for fallen lumps of rock that might contain fossils, but didn't find anything of interest so then headed back to camp, picking up the flags as I went. It snowed lightly the whole time I was away, causing visibility across the four-kilometer width of the glacier to fluctuate from poor to bad.

I spent rest of the afternoon diligently reading papers on the geology of the region and finishing my novel, *The Miko*, by Eric Van Lustbader. It was an easy day, but frustrating in having to just wait around until the weather fined up. I had a clear view of the mountains all around us and knew that much of this exposure had never been searched for fossils. It was virgin paleontological ground, the best kind.

Damn, I thought to myself, as I peered out of the tent next morning. It was still snowing. Visibility had worsened. We could not move on so were forced to remain there and wait it out until the weather cleared. The temperature was unseasonably quite warm at 21°F, although we were quite aware of the fact that warmer temperatures often heralded blizzards. I spent that morning reading Leon Uris' *Mitla Pass* and doing very little else. It's amazing how many good books you get to read while tent bound. Occasionally, we ventured out for short walks around the camp to stretch our legs, but it was an all pervading gloomy white-grey outside, a perfect kind of bright hazy light that would be ideal for getting snow blindness if you didn't wear your sunglasses.

For lunch that day we decided to cook a gourmet meal as we had all the time in the world. Later we sampled the experimental dehydrated yogurt that the field store supply man had asked us to test drive, but all agreed that it was pretty bad (far too watery). After dinner we clambered around the radio for our evening's entertainment. We heard news from Scott Base, passed on Christmas messages to be forwarded to the other groups in the field and dictated telegrams to be sent to our families back home.

Once more we awoke despondently next morning to the quiet sounds of snow falling on the tent. After our radio schedule to Scott Base at 8:00 A.M. most of us went back to sleep, tired largely because without any physical activity we were not sleeping very well each night. It didn't take much of this before our daily routine became out of sync with regular activity. For example, we had breakfast about noon, then I spent the afternoon reading and skipped lunch. I decided that I would prepare a really superb evening meal that night, mainly to give me something challenging to do to pass the time.

That night I cooked my most ambitious Antarctic field dinner yet: Thai satay beef with khao padt (pronounced like "cow pat"), which is Thai for fried rice. I had worked several field seasons in Thailand over 1988-90 searching for fossils, during which time I had acquired a love of authentic Thai food. The thought of a spicy hot Thai meal cooked in a polar tent in the remote deep field of Antarctica was a delicious contrast. The beef satay was easy, but the fried rice took a hell of a lot of effort. The rice had to be first boiled (after snow was thawed for the water), then each grain dried with paper towels. I rehydrated the dried onion and peas in hot water, then patted them dry and fried them lightly in oil. Some egg powder had to be mixed with water and milk powder to make a thin "omelets" which was then cut into thick egg noodles. Finally, I fried the whole lot with plenty of chopped bacon, oil, and butter. We had our last can of beer each from the recent resupply to accompany the meal. Later I read in my sleeping bag until I'd finished my novel at about 12:30 A.M.

The weather was still bad when we woke sometime late next morning; it was Saturday, 21 December. It had been snowing all night but later during the day occasional patches of blue sky appeared above. I walked down to the blue ice and found that the freshly fallen snow was now over 30 centimeters deep in places. It was a very cold day, so we didn't venture outside much. Wind blew the new snow around like grains of sand in a desert storm.

I spent that morning sewing up the tatters of my inner gloves, then began reading Ken Follett's *Pillars of the Earth.* The rest of the day passed slowly and uneventfully, but later that night the weather seemed to be clearing up. We could see Mt. Metschel and Portal Mountain bathed in sunlight out on the Skelton Névé. I went for a quick run around outside about ten o'clock to warm myself up and stretch my legs. Shortly afterwards I was sitting up in the tent inside my double sleeping bags eating chocolate and sipping Twining's peppermint tea. I wrote sarcastically in my diary those immortal words of Robert Falcon Scott: "God this is an awful place!"

The following day, our hopes were once more dashed as it continued snowing with winds gusting to about 15 knots, creating very poor visibility. We couldn't move on. We were forced into passing yet another day waiting inside the tent.

After a light breakfast I studied a draft of a paper that Gavin Young, Alex Ritchie, and I had just submitted on the new fossil crossopterygian fishes of the Aztec Siltstone. This group contained the predatory lobe-finned fishes whose only living relative was the coelacanth, *Latimeria,* discovered alive off the coast of South Africa in 1938. Only recently, in late 1998, another population of these living fossil fishes had been discovered off the coast of Sulawesi, in Indonesia. The Devonian crossopterygian fishes of the Aztec Siltstone included one giant creature up to four meters long, which we named *Notorhizodon,* as well as smaller blunt-headed forms with very small eyes, like *Koharolepis.* This latter species belonged to an endemic group of lobefins that we named the canowindrids, after the genus *Canowindra,* coincidentally named after its locality, the town of Canowindra in New South Wales. The canowindrids have so far only been found in East Gondwana, from sites within Australia and Antarctica. When British paleontologist Keith

Thomson described the first member of this group in 1973, its relationships were obscure. It could be placed with at least two major groups of ancient lobefins (called Porolepiformes and Osteolepiformes). This puzzled me for some years and in 1985 I had the opportunity to study the original specimen and write a new description of it. This time I emphatically placed the beast with the osteolepiform fishes, the group on the direct lineage leading to the first tetrapods.

My placement of the group as a primitive kind of osteolepiform was soon after reinforced by my description of another canowindrid, which I named *Beelarongia*, from Mt. Howitt in central Victoria. This second genus was older and more primitive than the first. Finally, the discovery of an even older member down here in Antarctica, which was clearly the most primitive member of the family, cemented the group at the very base of the osteolepiform radiation, and demonstrated beyond doubt that the eastern side of Gondwana (Australia and Antarctica) was once home to some very peculiar endemic fishes.

As I read over the draft of the paper I kept thinking how any day now, when we could get out onto the mountains to collect, we were probably going to find more of these fabulous yet enigmatic fishes and I would have to amend the draft manuscript with details of all the new discoveries! Although we did find other good specimens of these fishes in the latter part of the trip, we didn't add this into the original paper because waiting for the preparation of the new specimens would have held up the publication for too long. However, we did include all the new locality information for these finds to make the monograph complete.

I noted in my diary that we had no chocolate that day. We had to go easy on the food rations from then on as we were running low on certain items. Originally, we had intended to be at this site for only a couple of days then head back to our depot at the head of the Deception Glacier. We had stowed away our special treats for Christmas Day on the other sledges which were then waiting for us about 30 kilometers to the north. The weather that evening was still intolerable, with no sign of clearing. I stayed up reading my novel until the wee hours of the next morning. Then, just before turning in to sleep, I poked my head outside the tent and was amazed to see clear blue sky! My hopes were now soaring that the next day would be fine and sunny so we could get on with our work.

I eagerly looked outside the next morning only to be sadly disappointed. Complete whiteout conditions prevailed. I couldn't even see the Boomerang Range four kilometers away from us. Our mood was very dismal. We were not talking very much amongst ourselves at that time, instead each of us seemed to be more introverted, passing time reading or listening with headphones to their own music. It had been a full week since we became stranded at this same spot.

Each day we made our scheduled radio contact with Scott Base. Sometimes the Kiwis relayed to us news items deemed to be of "world news interest." Most days this might be a cricket score, who won the last Rugby test, or maybe news about some major accident back home in New Zealand (like some lady's washing blowing off the clothesline). Then, one day between cricket scores they told me incidentally that Australia had a new Prime Minister. Bob Hawke had been "retired" and Paul Keating had taken his place. No details accompanied this historic message, and they then went on to tell me that this had actually happened some while ago. I suppose that at the time it just didn't seem to rate as "interesting news" to them, because the sports always came first!

That evening after dinner we all worked outside to clear away the snow from the skidoos and sledges and off the outside of the tents. We made sure everything was secured for the night in case a storm came. Sometime after dinner we played a game called "throw the tea bag into the clown's mouth." We had built a snowman and colored him with various food items and attempted to swing used tea bags into his mouth. Well, at least it passed the time for a short while.

The next morning was Christmas Eve. It started off as a clear, cold day but with moderate winds gusting from 15 to 20 knots. By 10:30 A.M. it started snowing once more. My job was to clear out the makeshift toilet, which had filled up with drifting snow overnight. Initially we thought we were only going to be camped at this place for a day or two, so we had not gone to any trouble to build an elaborate "loo." It was simply a wall of cut ice blocks piled up around the portable loo, barely enough of a structure to cut off the person using the amenity from the view of the tents. It felt very cold squatting over the outside ig-loo when snow was being blasted at you from the open Névé at 15 knots! I used sludgy water to patch the holes in the ig-loo walls, letting it freeze up over the gaps.

Despite shortages of certain items like instant coffee, sugar, and most of our milk powder, the food was lasting well. We estimated that we had a maximum of another eight days of food left, if rationed carefully. Our Skidoo fuel supply was fine, but we were low on kerosene for the stoves. We knew at this stage that we would have to make a dash for our depot within the next week at all costs, despite the weather, as once fuel for the stoves ran out we couldn't cook or make water for drinking. Our predicament was starting to worry us all a little. Brian, for want of a physical job to do, went out that day and built us a better-sheltered, underground loo.* It was our Aladdin's cave, complete with a toilet down the bottom of the icy stairwell.

Christmas Eve was actually our eighth day of being tent bound. This sarcastic little poem highlighted the mood we were all in at the time:

Tent bound, eighth day in a row
Because of the bloody snow
The worst of it is not our plight,
Least we have warmth and a bite
But that the rocks are covered
Their ageless secrets smothered
By that bloody snow.

After dinner that night we played cards (pontoon) for a while as we waited eagerly for our radio contact at 8:00 P.M., desperately hanging out to hear any news of a break in the weather. That evening I received a telex from Ken McNamara, my colleague at the Western Australian Museum, passing on some interesting news items from home. The skeleton of a very large *Diprotodon*, an extinct fossil marsupial somewhat resembling a wombat the size of a rhinoceros, had been discovered south of Karratha in the Pilbara region. A Western Australian

*Australian for "lavatory."

Museum field party, led by my friend Alex Baynes, had gone out with a team of volunteers to excavate it. I could just imagine them in temperatures of 113°F sweltering away with jackhammers and generators over the giant fossil bones, while I was here languishing in the –4°F range. It snowed throughout the rest of the night, further dashing our hopes of a move the next day.

On Christmas Day, 1991, I awoke to find nothing at all in my smelly woolen socks, and was very disappointed to find that the biscuits and orange drink I'd left outside the tent for Santa hadn't been touched. It was quite windy, gusting at around 20 knots with the temperature strangely warm at 21°F. Visibility was becoming better that day, although the sky still had 100 percent cloud cover, so ground definition remained very poor. After our 8:00 A.M. radio call we slept in until 10:30 A.M., then Brian made us a rare treat of percolated coffee using some of his own private supply of ground coffee beans that he'd been saving for the occasion. It looked like we would have to sit out the bad weather a bit longer, so we decided to make the best of Christmas Day, despite being unfestively low on food and almost completely out of grog supplies.

Some of the guys in the radio room back at Scott Base sang us a somewhat unmelodious but heartwarming rendition of "We Wish You a Merry Christmas" over the radio, and Brian and I retaliated with a blaring yet woeful chorus of "I Saw Mummy kissing Santa Claus." Fraka then came over to our tent and the three of us played cards for a few hours while Margaret was busily preparing our sumptuous Christmas dinner. My job was to make a cheesecake out of a packet mix for our dessert. It was an easy job. To set it you simply place the fry pan with the cake mixture outside the tent for a short while.

At about 8:00 P.M. we jovially commenced our Christmas Day feast. The only alcoholic beverages we had left was a wee nip of Bailey's Irish Cream each and a small 50-milliliter bottle of Dimple Scotch that my bottle shop man in Perth gave me just before I left Australia. He told me to take it to Antarctica for Christmas, so I stowed it away in my kitbag, protected inside one of my boots. It was the only bottle we had to share on this festive occasion. Carefully we rationed out about 13 milliliters each and then made a small toast for Christmas.

It somehow reminded me later of the strange combinations of Christmas drinks drunk by some earlier expeditioners. Bages and his men on Mawson's Southern Sledging Party held their Christmas celebrations a couple of days later on 27 December, and had some rather diabolical homemade "wine":

> There was a general recovery when the "wine" was produced, made from stewed raisins and primus alcohol; and the King was toasted with much gusto. At the first sip, to say the least, we were disappointed. The rule of "no-heel taps" nearly settled us, and quite a long interval and cigars, saved up for the occasion by Webb, were necessary before we could get courage enough to drink to the Other Sledging Parties and Our Supporting Party.

Our Christmas dinner was immensely enjoyed by all, despite the dire shortage of drinks. In a way I sort of felt that this was my atonement for indulging in such a lavish Christmas dinner on Scott Base three years earlier while Margaret and her colleagues were toughing it out on the Darwin Glacier. Still, it wasn't a bad meal at all. We started with packet chicken soup followed by apricot chicken with a light whisky sauce, served with peas and rice. Margaret had concocted the sauce using a little of the remaining Bailey's Irish Cream mixed with apricot jam and soaked apricot pieces. Then we had some crackers and cheese, sweets that had been saved for the occasion, and an apple each, the very last items of fresh food from the last re-supply.

I had brought a present for each of the other field members and they had also brought along little gifts which we gratefully exchanged with our goodwill all round. I opened a Christmas present that I'd brought with me from home and was delighted to receive a book from my parents-in-law by ABC* journalist and Antarctic historian Tim Bowden, entitled *Antarctica and Back in 60 Days*. We all concealed our chronic homesickness well that day. A few tears were shed by all of us in our own way, in our own space, when the others weren't meant to be looking. I only had to listen to my tape from home and could barely hold back my emotions. Thoughts of Donna and the kids having Christmas with all the family in Melbourne filled me with great joy, closely followed by the sadness of realizing just how far away from

---

*Australian Broadcasting Company.

them I was at that moment. We joked that day about how we considered ourselves to be the most isolated bunch of humans on the planet to be celebrating Christmas.

Margaret and Fraka had adorned their tent with silver and red foil and some colorful paper hangings. Christmas cards were strung up, and we all pulled Christmas crackers and listened to the more melodious parts of my Christmas tape from home. Finally we sang all the carols we could remember until about 11:20 P.M., then turned in to bed, sober, but in very good spirits.

The next day the sky was clear and the sun shone brightly, but the winds were noticeably stronger, gusting at around 30 knots or more. It was very warm at 21°F. We were indecisive about moving, but due to our pressing circumstances decided that we would attempt to get to our depot at all costs, so reluctantly we packed up the camp in the strong winds. We left the tents up until the very last minute in case the weather suddenly took a turn for the worse. Finally, just before noon we pulled down the tents, lashed them to our sledges and slowly headed off towards Alligator Peak, following north up the front of the Boomerang Range. I noted that the strong winds were blowing the snow off the mountains in big white clouds, exposing the rocks! This was a good sign for collecting fossils.

It was a hard day of traveling as the sastrugi were jagged and high, and the ground definition was very poor as everything was draped with the newly fallen snow of the last nine days. Sledging was made difficult by many unexpected sudden crashes as we went blindly over hidden depressions. The winds blew between 20 and 40 knots constantly, sometimes much stronger in fierce gusts. By about 2:15 P.M. we had traveled 17 kilometers to a point about four kilometers or so out from the prominent saw-toothed ridge of dark volcanic rock known as Alligator Ridge. It was so named because it looks very much like the snout of an alligator in profile.

We pitched camp and ate a hearty dinner. Our depot was still about ten kilometers to the north of us, but there were good exposures of fossil-bearing Aztec rocks here so we were all keen to get some work done, despite being low on food and fuel. Strong winds howled all night between 40 and 50 knots.

At least, finally, we have moved from that accursed spot that trapped us for ten days, I thought to myself as I lay snuggled up in my sleeping bags that night. I was looking forward to searching for fossils again the next day, weather permitting. We planned to remain there and work on Alligator Peak if the weather improved. However, if the weather remained bad and we became desperately short of fuel and supplies, at least we could attempt a dash to our depot.

22
On the Snout of the Alligator

It was so much simpler—so much more normal—
to lay everything to an outbreak of madness on the
part of some of Lake's party. From the look of
things that demon wind must have been enough to
drive any man mad in the midst of this center of
all earthly mystery and desolation.
—H.P. Lovecraft

This passage is one that sums up well how Antarctic explorers often feel about the wind, which is enough to drive a person to the brink of insanity after weeks of incessant howling gales. For us, the wind was both evil and a blessing, at times uncomfortably cold and cruel, yet at other times blasting away the snow covering our precious fossil-bearing rocks, exposing the timeless treasures we had ventured down to this inhuman landscape to find.

On the first day after the weather mercifully eased up, we headed straight in towards the cirque enclosed by the Alligator's snout, a site formerly visited by early VUWAE teams in the 1970s and logged on their maps as simply "Section number 19." We hunted around on the low outcrops of Aztec Siltstone here but didn't find any well-preserved fish fossils, just a few scraps here and there. After lunch we decided to go out to Section 20, which involved climbing up a steep ridge leading to the summit of the mountain, Alligator Peak, which was just over 2000 meters high. It was a very productive afternoon up there as we found fish fossils in several different sedimentary layers. At the top of this section there were greenish calcareous nodules, reminiscent of "palaeosols"—the remains of ancient soil horizons. These are good indicators of ancient flood plain environments.

That evening we listened to the radio for news from Scott Base, then read and chatted in our sleeping bags until about midnight. I recorded some interesting trivia in my diary about how Brian and I were rambling on about a devious way to keep meat fresh in the Antarctic. He came up with the idea of taking a live pig on skis and towing it behind the sledges! Wrapped up well in thermal clothing, one could simply take its legs off one by one as was required for roasts, eventually leaving it on just two legs on the same side (thus it only needs one ski), or without legs the pig's body could be towed upon a single ski. This inane conversation was inspired by the fact that we still had about 50 packets of bacon left at the depot waiting for us! Maybe we were just going a little silly. At that time any conversation that inspired spontaneous humor was always brought up again and again until the theme was milked dry.

It resumed snowing the next morning, giving us a hazy view of the Alligator Ridge less than five kilometers away. Once more we could do nothing but sit and wait for the snow to clear before being able to get out onto the outcrops. We were quite low on food, and our depot was about twelve kilometers away as the skua flies. However, between it and us there was a large crevasse field that extended a long way out from the snout of Alligator Ridge, so when the time came to move on we planned to take a much longer but, we hoped, safer route to reach the depot.

Later that day it cleared so we climbed straight up Alligator Peak to search for more fossils, ignoring the annoying winds. We bundled up in our heaviest clothing, packed lots of food and drinks, and ascended the razorback ridge known as Section 21. We soon discovered that it had several good fossil fish sites. At one spot I found some well-layered black shale with very nicely preserved impressions of armored fish plates that stood out as whitish-green patches in strong contrast to the dark rock. Occasional white flecks in the rock revealed themselves under a hand lens as being the tiny scales of jawless fishes called thelodonts. Thelodont scales are amazingly like the teeth of later vertebrates in that they have a crown made of sculptured dentine over a bony base with a large pulp cavity in it. By invading the mouth regions, scales like these would eventually evolve into the first teeth in jawed fishes. Thelodont scales are also widely used around the world for dating sedimentary sequences, so their presence here was an important discovery, as previously they were only known to occur down at the base of the Aztec Siltstone. Therefore this find suggested that we were probably low down in the geological section.

Several years later I wrote a paper on those few fish plates collected from this site on that day. I named a new genus of extinct fish, called *Boomeraspis*, meaning "the shield from the Boomerang Range." *Boomeraspis* was an interesting form of placoderm intermediate between the *Groenlandaspis* species commonly found in Australia and Antarctica, and the more primitive phlyctaenioid arthrodires well known from the Northern Hemisphere sites in Europe, Spitzbergen, and North America. In simple terms, the presence of my new fish here in Antarctica was a strong indication that the genus *Groenlandaspis* may well have had its origins in East Gondwana. Such finds are important to refining the biostratigraphic correlations across Gondwana countries as *Groenlandaspis* is otherwise restricted to just the very latest part of the Late Devonian (the top of the Famennian stage) in the northern hemisphere, but occurs much earlier in time in both Australia and Antarctica (in the Givetian and Frasnian stages that precede the Famennian). A few years ago I was working in South Africa in the Barrydale-Ladismith region and became the first person to recognize *Groenlandaspis* from that continent and, guess what? It also occurred in Middle Devonian rocks, precisely the same age as the Aztec Siltstone of Antarctica and the Mt. Howitt deposits of central Victoria. Once more the Gondwana link (Australia-Antarctica-Africa) had proven to be a reliable predictor of the earlier chronological appearance of this important fossil fish.

At the top of Alligator Peak I found more *Groenlandaspis* and *Bothriolepis* plates, and Margaret found the trace fossil *Beaconites* occurring in the same layers. I was quite excited by the discoveries that afternoon, even though at the time I had no idea that I had actually bagged a new genus of fossil fish that day. Although for most of the day it was sunny and clear, strong winds blasted us from time to time so we had to don our warmest gear. We loaded up our packs with as much as we could carry then eventually headed down the ridge towards our camp, and were ready for dinner by about 8:00 P.M. After our meal the wind dropped off and I lazed around outside in the sunshine, reading my new Tim Bowden book while draped over a sledge.

About 10:30 P.M. that night it suddenly hit us all simultaneously that the wind had calmed down and the weather had fined up. Despite the time, and weariness from working a long day up on the mountain, there was a unanimous decision to make a midnight dash for our depot!

23

Long Day's Journey into Night

And on such a day I have seen the sky shatter like a broken goblet, and dissolve into iridescent tipsy fragments—ice crystals falling across the face of the sun. And once in the golden downpour a slender column of platinum leaped up from the horizon, clean through the sun's core; a second luminescent shadow formed horizontally through the sun, making a perfect cross. Presently two miniature suns, green and yellow in color, flipped simultaneously to the ends of each arm. These are parahelia, the most dramatic of all refraction phenomena; nothing is lovelier.

—Richard Byrd

The beautiful description of Richard Byrd's experience of seeing the parahelia on 13 April 1934 highlights the almost spiritual, transcendental kind of beauty that certain moments in Antarctica can convey, touching a person's inner soul with a subtle sense of oneness with the universe. I recall a similar experience when we traveled over the glacier throughout the night, in totally peaceful, still conditions.

We were ready to head off at midnight. The continuous fall of light snow from the past ten days had draped the entire landscape with a thick white blanket. Coupled with the cloudy weather and diffuse daylight, this made the ground definition rather poor, so that bumps and depressions could barely be discerned. Nonetheless, we had to move northwards to our depot of food and fuel. We started off by heading out a long way into the center of the Skelton Névé, skirting well round the protruding snout of Alligator Ridge with its accompanying large crevasse field. It was fairly easy going from that point onwards.

The sun hung low in the sky, casting an almost hazy purplish hue over the newly fallen snow that mantled the rocks and sastrugi, smoothing the landscape with its gentle undulations and subtle reflections. The whole scene was bathed in an almost surreal glow. There was also an eerie absence of any wind noise after the almost incessant gales of the last ten days. The only sounds I could hear were the rhythmic swish-swoosh of the sledge runners below me and the distant drone of the Skidoo engine, about 50 meters ahead, as we moved solemnly along. My mind was spinning, I was singing old songs aloud to myself and my heart was full of joyous tranquility. I felt incredibly lucky to have experienced these moments, at that time and in that special place, and I now carry them with me always.

Byrd relates an almost spiritual experience he had one night during his winter alone in 1934, where the majestic chords of Beethoven's Fifth Symphony merged with the visual splendor of the aurora in a twilight evening: "The music and night became one; and I told myself that all beauty was akin and sprang from the same substance. I recalled a gallant, unselfish act that was of the same essence as the music and the aurora."

The surface of the Névé was flat and smooth from the newly fallen snow. At one stage the going was so easy that I dared to climb up onto the back of the sledge, right on top of the gear and sit there with my legs stretched out. I lay down for a brief moment, my eyes staring upwards at the clear night sky, my thoughts rapidly wandering away from reality. Brian towed me along, but after a few moments I became aware that my lack of concentration could prove to be dangerous, so I straightened up to watch the path of our journey carefully.

We reached our depot close to 1:00 A.M., after just under an hour of wonderful traveling. The two sledges at the depot were covered in snow from the almost continual falls of the last two weeks. It took us about half an hour to re-hitch the second sledge to each of the skidoos, then we decided that while conditions were so good we should attempt a dash towards our next fossil site to set up a camp at Mt. Metschel. At the west end of Mt. Metschel very strong winds were blasting down off the crevassed slopes. The sledges had to be steered hard and worked constantly to keep them on track with the skidoos and to prevent overturns.

We reached Mt. Metschel at about 3:40 A.M. It was then starting to get very cold and blustery so we quickly searched for a suitable place to pitch our tents. Unfortunately, there were no obvious campsites with reasonable snow cover for the winds had stripped it all away, exposing just jagged blue ice. In the end we decided to get out the ice screws and strike camp upon the hard glacier. This involved having to hammer in the ice screws to secure down the corners of the tents and instead of shoveling snow over the extended tent flaps we packed them with heavy rocks.

About 5:00 A.M., exhausted from the full day's work, long night's journey and having pitched camp on the blue ice, all four of us gathered in our tent for coffee and to finally partake of some of the delights of the new food boxes, bristling with fresh supplies of chocolate. A special luxury we sampled that morning was the homemade cake originally destined to have been one of Margaret's special surprises for Christmas Day. It was excellent. We were all feeling pretty pleased with the night's work, so we opened a bottle of Southern Comfort and passed it round with the cake. Soon after Margaret and Fraka stumbled off to their tent to sleep.

An hour or so later, I heard Fraka and Margaret laughing and giggling in the next tent. Brian and I had almost finished off the bottle of Southern Comfort by then. The standard joke about having it "on the rocks" had worn a bit thin by that stage. Brian was chatting more about the merits of taking live pigs along on NZARP deep field expeditions, a subject now dear to our hearts. Basically, after having no alcoholic drink supplies for the previous week, it was no wonder we were all quite merry on that occasion. Mt. Metschel looked very beautiful from our tents with its alternating olive green and red bands of sedimentary rock below a thick black dolerite peak.

Brian and I had another "nightcap," demolishing the bottle of Southern Comfort, then chatted until 7:30 A.M. when we contacted Scott Base with our new position. There had been no response to our earlier call at 4:30 A.M. Then, having done this last duty, I slept deeply until the early afternoon. On waking I fired up the primus and had a simple lunch of porridge and coffee. I then got my gear ready to go out and search for fossils.

That afternoon I wandered out alone to explore Mt. Metschel as the others slept inside their tents. Large *Bothriolepis* plates were scattered around through the rocks with abundant crossopterygian bones and scales in the top units. I climbed over massive flat pavements of grey silty sandstone, making the whole cliffside seem like the world's largest amphitheatre. After a few hours of collecting, I sauntered back to camp for dinner. We were all in bed by 10:00 P.M. I noticed how exhausted we sometimes became after very long days of sledging, as it would sometimes take us two days to fully recover from the ordeal.

The next day was clear with light winds. We decided to examine two main outcrops at Mt. Metschel. The first, Section 13B, as named by the previous field party, is situated about 500 meters north of Section 13A, the site where I had collected some nice specimens the previous day. After looking thoroughly around these sites we then drove around to the southern end of the mountain and walked up some beautifully terraced exposures of reddish-brown mudstone. This site was amazing in that we had a small series of flat rocky platforms surrounded by a huge cliff of jagged ice, forming an icy cirque towering about a hundred meters higher than the rocks, somewhat akin to a hole carved into the ice with a low rocky bluff exposed in the middle. It was well sheltered from the wind and received some sunshine in the early afternoon, so made for pleasant working conditions.

One of the most spectacular finds of the whole trip was made here on the very top layer of exposed mudstone. I found a mass of bright orange scales and bones that bore a shining "cosmine" surface. Cosmine is a special type of dentinous tissue unique to certain types of extinct lobe-finned fishes. It has little flask-shaped cavities in it that are interconnected by tubules below a dentine layer. What all this means is that the highly porous nature of this surface tissue is interconnected with the fish's sensory-line system, possibly making the whole exterior surface of the fish into an electro-receptive organ. I knew just from the shape of these scales and their glimmering surface that they came from a very large predatory lobe-finned fish, maybe two to three meters long.

The strangest coincidence occurred some years later, as I was rummaging though the fossil collections of the Australian Museum in Sydney. I came across a specimen collected by Alex Ritchie from the exact same site at Mt. Metschel. I saw large chunks of bones, scales, and teeth preserved in the same kind of brownish mudstone, and all had the same characteristic bright orange color. I have no doubt that I had collected more of the same beast that was exposed at the surface when Alex was there twenty years before me. The weathering action caused by water thawing on the dark rock and freezing in the cracks had fractured the rock into hundreds of small pieces, exposing more of the giant fish, but as I collected all the fragments I could find, I think there is a good chance of one day piecing it all together and reconstructing its huge fossil skull by using the Australian Museum material to complete the jigsaw. Cross-sections in the rock show large teeth and jawbones, indicating it was an osteolepiform fish, the same group which eventually gave rise to the ancestors of the first land animals, the tetrapods.

The 1970-71 VUWAE 15 expedition had collected a large cosmine-covered skull from Mt. Crean, which Gavin Young, Alex Ritchie, and I described in 1992 as a new genus, *Koharolepis*, meaning "shining scales" after the Maori word *kohara*. As explained previously, this beast was one of the canowindrid family, an endemic group of extinct fishes known only from East Gondwana. The new specimen from Mt. Metschel is a much larger fish, having scales nearly twice the size of the Mt. Crean beast. It is exciting material and in 1990 I published a paper suggesting that the first tetrapods may have originated in East Gondwana. I put this hypothesis on the evidence that we had remains of Devonian amphibians in New South Wales, based on a jaw named as *Metaxygnathus* found near Forbes; and we had lobe-finned fishes that were more primitive than anywhere else (as argued more fully in our paper of 1992).

Lately, though, this idea has been losing ground as many remarkable new discoveries from Europe and North America are showing that the true tetrapods did appear earlier in the Northern Hemisphere, and that fishes more intermediate between amphibians and osteolepiforms also existed there: a group called the panderichthyids. The clincher came recently when my colleague Dr. Per Ahlberg of the Natural History Museum in London found the remains of a very primitive fish-like amphibian that he named *Elginerpeton*, after its locality near Elgin in Scotland. The picture is now emerging that tetrapods most likely evolved from lobe-finned panderichthyid fishes by the end of the Frasnian Stage, about 370 million years ago, in the northern hemisphere.

However, the early radiation of the lobe-finned fishes may still be a Gondwana phenomenon. Gavin Young has suggested that there is good evidence for a faunal interchange event between the northern hemisphere landmasses Euramerica/Asia and Gondwana at the end of the Middle Devonian, about 380 million years ago. This could have allowed the primitive lobe-finned fishes from Gondwana to get into Euramerica and from these ancestral forms the more specialized panderichthyids could have arisen. Any piece of the evolutionary puzzle that sheds new light on the mystery of tetrapod origins is an important scientific find, so I was more than pleased to collect the skull of another early lobe-finned fish, possibly one that will turn out to be a new genus, from the top of that low rocky platform on the southern edge of Mt. Metschel.

For dinner that evening Brian cooked an unusual chicken recipe, peppermint chicken à la Staite, a delicate dish that resulted from the fact he had inadvertently used the frying pan in which I had spat out my toothpaste water that morning. We never told Margaret or Fraka where the unique flavor of the dish came from. Actually it wasn't too bad, and for the benefit of doubters, I have included the details of the recipe in the appendix at the back of this book. Just make sure you have a good brand of toothpaste or it could taste awful!

It certainly wasn't the first time that honest mistakes had been made and not recognized in the long history of Antarctic field cookery. Apsley Cherry-Garrard notes how on the depot journey of Scott's 1912 expedition he was tired and made a mistake mixing up the cocoa: "It was dark, and I mistook a small bag of curry powder for the cocoa bag, and made cocoa with that, mixed with sugar; Crean drank his right down before discovering anything was wrong."

We heard a rumor that evening from Scott Base that VXE-6 Squadron planned to pull us out on 10 January, so if this eventuated we would have only two weeks left to reach the remaining localities where we wanted to work. Our original plan was to sledge up to Mt. Fleming, just inland from the Dry Valleys where I had explored in 1988, and be picked up from there. This would position us within easy helo range of McMurdo Base, but it seemed unlikely now that we could get that far by the suggested pick-up date. Mt. Fleming was about 140 kilometers to the north of our current position, a long way to travel in unpredictable weather conditions.

I received a telegram that day from my family, read out to me over the radio from Scott Base. My daughter Sarah was excited because she was going to see *Phantom of the Opera* in Melbourne on 10 January. I also learned that recently Sarah and my son Peter had lost a tooth each. I always felt that when Scott Base announced there was news from home it might just be bad news—if that was the case I'd be powerless to return home at once to be with my family. So, any good news from home was always uplifting, no matter how trivial it might seem in retrospect.

The last day of 1991 we packed up camp and moved to the Portal, 24 kilometers away at the head of the Skelton Névé. It was fairly easy going over the thick mantle of recently fallen snow. As we left Mt. Metschel we found a damaged polar tent, perhaps one lost by an earlier VUWAE field party.

We arrived around 2:30 P.M. and immediately selected a camp site in the shadow of the Portal Mountain, about 200 meters away from the impressive cliff faces draped with heavy snow. The Portal was so named as it forms one of the major gateways to the polar plateau up the head of the Skelton Glacier. It is a towering 2556 meters high and is mostly composed of layered sedimentary rocks with only small amounts of black dolerite rock.

There was a small depression, of which I didn't think much at the time, between the mountain and us. Later we would discover that the whole area was riddled with crevasses. Our campsite was chosen on a slight hummock surrounded by a thick snow cover all around it. It was lucky that we arrived early in the afternoon, as snow began falling the moment we started to pitch the tents. By 3:30 P.M. we had finished but had to wait yet again until the snow stopped before we could try and get out onto the outcrops to search the fossil-bearing horizons. I was very excited at finally getting here as the Portal was one of the best fossil fish sites known from any of the Aztec Siltstone outcrops.

On the 1970-71 VUWAE expedition Gavin Young and Alex Ritchie had collected a large number of very well-preserved specimens here, including some complete head shields of placoderms and some very large fossil shark's teeth. A number of smaller undescribed placoderm plates from the site were tantalizing in that they belonged to family groups poorly known in Australia, such as the phlyctaenioid arthrodires. These were little armor-plated fishes with wide triangular pectoral fins guarded by flaring spines. The significance of these fishes has to do with their evolutionary position as the most likely ancestral group to the later groenlandaspid and phyllolepid placoderms. Only one species of this entire group, very diverse in northern hemisphere countries, had been discovered in Australia previously.

One very interesting placoderm specimen was collected from the mountains of this region by the 1955 geological sledging expeditions led by New Zealanders Bernie Gunn and Guyon Warren. It was a tiny partial skull about four centimeters long that they sent to Dr. Errol White of the Natural History Museum in London. White named this fish *Antarctaspis mcmurdoensis.* This intriguing little skull was actually a missing link for us placodermophiles as it showed characters intermediate between the regular arthrodires and the peculiar flattened phyllolepids. Further scientific papers by Gavin Young have alluded to East Gondwana as being the center of origin for the phyllolepid placoderms, so I was very eager to search here for more remains of this little beast and any of its close relatives. So, having to wait once more to get up onto the outcrop was frustrating for me to say the least.

That evening it continued snowing heavily as the night grew colder. Not only was it Margaret's 50th birthday but it was also New Year's Eve, so we had more than ample cause for celebration and enough grog supplies to indulge our whims. We feasted on a large turkey that was originally meant for our Christmas Day dinner. We then mixed up a bottle of our patented "Deception Irish Cream" and finished that along with a bottle of Robard & Butler's Artillery Port. I often wondered why it bore this name and thought that it must be because the empty bottles were used for target practice by the artillery.

Needless to say we were all in high spirits that night, and played "pass the pigs" and "Black Maria" until quite late. At midnight we went outside, held hands in a circle and loudly sang "Auld Lang Syne" while dancing around in a circle. Then we had a snowball fight, wrestled around on the ground, and carried on with play fights for a short while until we were all quite exhausted.

Inside the tent we listened eagerly to the radio for messages from Scott Base and other field parties as the time approached midnight. We were greeted by many New Year messages, including one from our old friend Garth Falloon, one of the surveyors just back from the McMurdo Ice Dome, who arrived in at the base just before midnight!

We also heard from Paul Fitzgerald's group (K5076), who had finally been successfully pulled out of northern Victoria Land. They had only one beef: they had to leave all their rock samples and gear behind in the field. They told us optimistically that VXE-6 Squadron was planning to go back and eventually retrieve their things. We merrily chatted to everyone on the radio till around 1:00 A.M., then had some more to drink, and finally turned in to bed about an hour later. We all slept like babies as the New Year settled upon us.

That was the end of 1991 for us. But 1992 had already begun, with promises of exciting discoveries up on the Portal once the weather became favorable.

Little did I know that 1 January 1992 would come dangerously close to being my last day alive.

24
Skating Away on the Thin Ice of a New Year

So it happened that as I fell through into the crevasse the thought "so this is the end" blazed up in my mind, for it was to be expected that in the next moment the sledge would follow through, crash on my head and all go to the unseen bottom. But the unexpected happened and the sledge held, the deep snow acting as a brake.

—Douglas Mawson

New Year's Day, 1992. I had been in the field in Antarctica since 15 November, just over seven weeks. It seemed like only yesterday that we had been shoved out of the backside of a ski-equipped Hercules on the Darwin Glacier, yet we had all gained a lot of confidence at Antarctic travel and mountain exploration. A little too much confidence can be a dangerous thing, though, and I was the least experienced member of the party.

I woke up for the 7:00 A.M. radio schedule and heard Brian say that snow had been falling most of the night and that conditions were not good for working. So I got up, put on a single-piece bunny suit and my Sorrell boots, and hesitantly crawled out into the chill morning to check out the weather. Back inside the tent a few minutes later I snuggled right down into the two layers of down sleeping bag, temporarily forgetting all about my plan to search for fossils, and slept deeply until about 11:00 A.M. Brian and I had a hearty breakfast of porridge, bacon, and toast. I enjoyed soaking up all the bacon fat with the bread, and I think at the time my body craved for fats of any kind. I washed the food down with about three cups of tea and coffee. Brian then lay back inside his double sleeping bag and began reading a book. I felt restless. We were so close to those famous fossil sites that I just wanted to get out there and collect, no matter what.

After the recent ten-day period in the southern Boomerang Range of just biding time for the weather to clear I had clearly had enough of waiting. We seemed to be always waiting for something in Antarctica. Waiting for a recon flight. Waiting for a put-in flight. Waiting for a re-supply. Waiting for a letter or telegram from home. Waiting for bloody Godot. But waiting for good weather was the most common pastime when in a remote deep field situation. On average an expedition expects to lose about one day in four due to bad weather. We had spent nearly two months camping out in the depths of southern Victoria Land, and had covered a distance of nearly 500 kilometers in Alpine skidoos and Tamworth sledges. At this stage the expedition had only about a week to go before pick-up day when our cargo cult would save us; a great silver and orange C-130 Hercules would come for us.

The conversation I had with Brian that morning went something like this:

"It doesn't look too bad out there now, Brian," I said casually. It had stopped snowing. I could see the outcrops of fossil-rich rocks only about 200 meters away, tempting me to come to them and partake of their earthly pleasures. I could just smell the fossils out there! My hand itched for my geology hammer.

"The snow's pretty thick out there now," replied Brian. "I spoke with Margaret and Fraka and they've decided to stay inside for the day."

"Would it be okay if I go out for a short walk and just check out that rocky outcrop nearest the camp?" I looked at Brian with wide puppy-like eyes as if to say "I've got plenty of experience now, nothing can happen to me."

Brian poked his head out of the tent and had a good look around. He consented, but warned me to take care as the weather could turn bad again at a moment's notice. Bad weather in Antarctica is not the same as, say, getting caught in the rain back home. It must always be taken seriously.

"Don't worry," I said, "I'll be extra careful. If I'm not back by 6:00 P.M. at the latest, send out a search party."

The last remark was not a joke.

After having poured the rest of the hot water left over from breakfast into my thermos with a packet of orange fruit drink added, I grabbed a large 250 g block of chocolate from the food box, a packet of army biscuits, and some small muesli bars and put them into my backpack. There was still some cheese, dry biscuits, and dried apples in my pack from lunch a few days ago, as food rarely spoils in the cold Antarctic climate. This meant that I had enough food and drink to tide me over in the extreme case of bad weather preventing me from getting back on time, forcing shelter for the night on the mountain. I then went over to Margaret and Fraka's tent and said a quick hello and told them I was "just going over for a quick squiz at the rocks and wouldn't be very long." On reflection this did seem a bit similar to the last spoken words of Lawrence "Titus" Oates from Scott's expedition: "I am just going outside and may be some time."

I grabbed an ice pick off the sledge and tramped away from the camp towards the face of Portal Mountain. Almost immediately, after about twenty steps, when I had gone beyond the raised hummock on which we were camped, I noticed the snow getting noticeably deeper as I pushed forwards in snow up to my mid-thigh. I kept vigorously making my way through the snowdrift, and the effort of it actually made me sweat despite the extreme cold. I stopped and looked back. The camp appeared kind of unearthly from a short distance away—two tiny bright yellow pyramid tents, four wooden sledges, and two shining orange toboggans in a vast sea of white, dwarfed by the awesome, dark brown brooding mountains.

Away from camp one hears nothing but the eerie silence of the winds, the squeaking of fresh snow under foot, and an occasional loud cracking noise, like an explosion, the result of the subterranean grinding blue ice of the glacier as it moves. Nikki Gemmel expressed this beautifully in her novel, *Shiver*: "The ice feels like Waterford, smooth and diamond hard. And then to hear the wind funneling around it, to hear the ice crack and pop and then be silent. Oh my sweetness! Just to listen to the nothingness, punctuated by the ice splintering. The ice is alive with sound, but you have to be still to hear it."

Well, it certainly has been snowing a lot here lately, I thought to myself. I was struggling to move in the waist deep snow. Nonetheless, I forged steadily forwards and was pleased to see the white and greenish-grey layers of ancient sandstone and shale getting closer. Who knows what treasures could have awaited me when I reached that outcrop? Perhaps some bizarre new species of fossil fish, or a new discovery that would pinpoint Antarctica as an ancient center of a new evolutionary explosion. Such ideas had already formed in my mind before I came to Antarctica and were quickly cementing as the real evidence we had found so far supported the theories.

I was then about 100 meters from camp. The snow was still quite deep and sounded very soft, almost hollow in places. I was dressed in my heavy-weather gear, a full one-piece bright yellow bunny suit, two layers of gloves, and Sorrell boots over two pairs of woolen socks. The struggle through the thick snow was very tiring. Often doing really simple things in the field in Antarctica can be exhausting thanks to the thick layers of clothing and the bitterly cold winds. The act of walking through this thick snow wearing my heaviest layers of clothing and carrying a fully laden backpack was just such a situation.

Suddenly my foot stepped right through the ground, and the other foot gave way also. I felt the horrible sensation of falling, with nothing below me, and instinctively thrashed around to suddenly break the fall as my large pack wedged itself deep in the snow. My feet wiggled in mid-air atop of a bottomless chasm. I frantically writhed sideways, rolling over to the right, snow and ice in my face and hair. Icy breath panting furiously! I was away from the crevasse. Safe. I got up and moved closer to the messed-up area of snow. A gaping dark blue hole with no visible bottom now existed where I had broken through.

My pulse was racing like Phar Lap,* but I quickly recovered from the ordeal. In the ensuing minutes I didn't think straight, couldn't really let the emotional weight of the moment sink in. Instead, I had just

*A very famous Australian racehorse.

one thought, to get away from the crevasse and onto the safety of the nearby rocks. Crazy thoughts about life, death and my family at home raced through my mind. Perhaps I was desperately trying to dispel the whole incident from my mind, put it all behind me, as I dragged myself through the waist-deep snow as fast as I could move towards the rocks.

I walked well away from the line of possible crevasses that usually border rock faces. Carefully treading the ground with each step I was soon safely on the solid sandstone ledge. I sat there, catching my breath, panting clouds of foggy breath into the stagnant chilled air, and quietly reflected on the whole experience for a few minutes. Then I had a drink from my thermos and a comforting chunk of chocolate.

Shit, that was nearly it, I thought again. Almost my life over then and there. I thought longingly of my wife and three gorgeous kids back home, how for a brief moment I came close to never seeing them ever again. Then a wave of emotion hit me like a train out of a tunnel and I broke down. I cried. My tears fell onto the weather-scarred Devonian sandstones, seeped into the ancient pores, and froze on the spot. I suppose I must have lost it for about ten or fifteen minutes, before I started to look around me, take in the spectacular view and slowly start to regain my composure.

After I had pulled myself together there was only one thing to do—I had risked my life to get to these rocks so I was damn well going to have a look around for fossils. I pushed on. Besides, I thought to myself, up on the mountain I knew I'd be safe, well away from the hidden crevasses. I slowly ascended the steep rocky slopes of the mountain. Due to the continuous heavy snowfall of the last two weeks, which caused our hold-up in the Southern Boomerang Range, these rocky platforms were covered by a thick, precipitous build-up of snow. After a short climb up the frozen rocky face I found myself on a small, rubbly outcrop of dark rock. I immediately recognized these rocks as dolerite, the black volcanic rock. Yet over to my left, about fifty meters away, was a definite outcrop of layered sedimentary rock showing the characteristic bands of dark, fissile shale resting on a three-meter bluff of grey sandstone. I felt that if I could get there I was almost certainly going to find the famous fossil beds first examined by Gavin Young and Alex Ritchie 20 years ago.

There was only one problem. Between the outcrop and me was a thick snow bank, and this was very steeply inclined. I decided I'd better put on my crampons over my boots to grip the snow better, then tentatively stepped out onto the steep slope. By stepping down hard I actually made a stable foothold, and then used my ice axe to probe the next foothold. The snow was as deep as my chest in places, and it felt just a little scary not to be able to see what surface I was walking over. By plowing forwards, almost swimming though the deep drift, I was soon almost at the next outcrop.

Just then I inadvertently noticed a few small pieces of snow bouncing off the ground next to me, whizzing past at frantic speed. I turned and looked up the steep slope above and saw a wall of snow bearing down at me. Instinctively I turned my back to the avalanche and covered my face. Shit! Someone's really got it in for me today, I thought. The falling snow hit me with such force that it knocked me forwards and I could feel the snow piling up rapidly all around my tumbling body. After a few seconds that seemed like an eternity, as my mind was full of adrenaline and morbid thoughts, the falling snow ceased, leaving me buried up to my neck. It must have looked really funny—a little white-faced, scared-shitless head poking up out of a mound of dumped snow, halfway up a treacherous mountain. As the snow had fallen recently, and hadn't had time to compact, it was actually fairly easy to dig my way out and plough the extra few meters to reach the safety of the rocks.

I then had a good long rest, sitting on a ledge high above the vast Skelton Névé. I could see the Boomerang Range with crystal clarity about 60 kilometers to the south, and our little camp only a few hundred meters below. I didn't want to move from that spot. I felt scared and vulnerable. My emotions were all over the place, but I could do nothing, so I just sat frozen to that spot. How trivial is man's presence in Antarctica, I thought to myself, and eventually summoned up the courage to push on.

I gingerly climbed down the rocky slope and somehow managed to find a few scrappy fish fossils in the lowermost layers. As I carefully traced my footsteps back to camp, I took great care to skirt well around the gaping hole where I'd almost dropped into the crevasse.

Finally I reached our tent at about 4:00 P.M. I told Brian about my two close shaves with the crevasse and the avalanche and he poured me a large glass of whisky. I spent the rest of the day in my sleeping bags, reading a little and occasionally writing down notes in my field book. I was quite unnerved and couldn't sleep easily. About 2:00 A.M., somehow, I drifted off to sleep.

The funny thing about that little foray onto the Portal on 1 January, which only took about three hours, was that the next day Brian carefully retraced my path, testing the ground with the crevasse probe to find a safe access route onto the rocks. He discovered that I had walked over seven crevasses, each with a thin snow bridge covering their almost bottomless chasms. By virtue of sheer good luck I had walked over all of them and only broken through one.

My experience of almost plummeting down a hidden crevasse recalls that of many of the Antarctic explorers, and some recent expeditioners. Admiral Richard Byrd's account of a similar experience sums it up nicely: "Then I had a horrible feeling of falling, and at the same time of being hurled sideways. Afterwards I could not remember hearing any sound. When my wits returned, I was sprawled out full length on the snow with one leg dangling over the side of an open crevasse." Later Byrd moves slowly away, and then checks out the crevasse, which is about a meter wide but very deep. He shines his torch down and says: "I could see no bottom."

Just another day in the Great White South. Ever since that day, whenever I've heard the old Jethro Tull song "Skating away on the Thin Ice of a New Day," I've thought of the Portal, and how on that day I skated away on the thin ice of a new year and, by a stroke of good luck, somehow survived to tell the tale.

25
Working on the Portal

We traveled for Science. Those three small embryos from Cape Crozier, that weight of fossils from Buckley Island, and that mass of material, less spectacular, but gathered just as carefully hour by hour in wind and drift, darkness and cold, were striven for in order that the world may have a little more knowledge, that it may build on what it knows instead of what it thinks.
—Apsley Cherry-Garrard

The horrific tale of endurance under atrocious conditions during the mid-winter journey to Cape Crozier by Wilson, Bowers, and Cherry-Garrard to collect three emperor penguin eggs was undoubtedly the inspiration for the title of Apsley Cherry-Garrard's book, *The Worst Journey in the World.* But, as the above passage indicates, their collection of scientific specimens was the justification, their *raison d'être* for keeping up the pace when the going got tough. The hardships endured by these early explorers were then considered to be a reasonable price to pay for advancing the scientific knowledge of the world, because at that time people had great faith that science would eventually solve some of the world's immediate problems. I feel in my heart that these days the world should embrace this concept and trust science rather than blame it for all its ailments. When I was working in Antarctica we had only one thought to drive us onwards and console us in our time of homesickness, that of discovery. Every site we explored had the potential to yield new specimens, new data, and new evidence to solve age-old mysteries of evolutionary science. So, we pressed on, regardless of past traumas.

The second day of the New Year was a clear, sunny morning with a temperature of 10°F. We were camped at an altitude of 1,590 meters in the shadow of the Transantarctic Mountains, which forged up skywards another kilometer or so above us.

Brian's newly flagged route gave us safe access to the base of the Portal. We scrambled up the mountain and worked all day on the snow-covered slopes exploring the wide, flat terraces of dark shale and reddish-brown mudstone. White flecks of fossil bone glistened in the sunshine all around us.

Amongst the many specimens we collected that day were more large, well-preserved fossil sharks' teeth. Some of these had two divergent cusps, which I would later describe and name in honor of our gallant leader, Margaret Bradshaw, as *Portalodus bradshawae* (meaning "Margaret's tooth from Portal"). Gavin Young and Alex Ritchie had collected similar teeth on the earlier VUWAE 15 expedition from the Portal. Gavin had tentatively identified them as *Xenacanthus.* However, a recent upsurge of new work on the xenacanthid shark teeth by Dr. Oliver Hampe of the Berlin Natural History Museum had enabled me to demonstrate that the Antarctic specimens were quite different. Oliver and I later described the internal structure and histology of these teeth, which has further shown that they are even more distinct from *Xenacanthus* and its allies than previously thought.

Portalodus has the largest teeth known of any Devonian shark. Whereas most sharks' teeth of this age are between a few millimeters to a centimeter long, the largest specimen of *Portalodus* is a full two centimeters in length, giving an estimated maximum size of this shark of about three meters. This clearly makes *Portalodus* one of the largest predators in the ancient river and lake systems of Antarctica. I can just imagine it cruising the bottom of the murky lake depths, flashing silver scales and white teeth as it flicks its tail and grabs an unsuspecting little *Bothriolepis.* It then chows down on yet another placoderm box lunch.

We ate our lunch that afternoon on a very steep slope overlooking the whole of the Skelton Névé, a sea of snow bathed in soft white mellow sunlight. It was a crystal clear day, showing the mountains rising out of the perfectly blanketed snow cover for many hundreds of kilometers. Down below us we could make out a very thin dark line wiggling through the snow from Mt. Metschel: it was our trail left by the sledge trains.

Although it was a mild afternoon, the Portal was so high that it kept us in shadow for most of the day, so it soon became quite cold on the mountain slopes. We had filled our backpacks with specimens by 7:00 P.M., then we headed down towards the camp.

I was a bit lost in my thoughts all that day, both simultaneously on a high from finding great fossils all around me, but also occasionally sinking into morose thoughts, dwelling on my previous day's bad experiences. I could see the crevasse field below with my footsteps wandering right through them, punctuated by a black hole in the middle where I had thrashed around on the snow. Still, the day passed quickly as I was so busy collecting specimens that most of the early evening I was obliged to write up field notes and carefully wrap fossils.

After dinner that evening we had some "kerosene cake," aptly named because some kerosene had leaked on it. It was still edible even though it smelt quite bad. We finished the meal with some of Brian's brewed coffee and a couple of nips of whisky each.

That night we played around outside for a while, pretending we were samurais and ninjas and so on. I recalled my childhood days of watching Shintaro the Samurai on TV and how I used to love playing at being ninjas, so we indulged ourselves by clowning around in the thick snow. The three of us attacked Margaret, who fell backwards laughing loudly into a depression in the snow. It was lots of fun and a much needed stress release for us all.

It was 14°F when we got up the next day. It had snowed constantly overnight and the sky remained overcast. We walked back to the lower section on the Portal. I found some fish fossils about a hundred meters up from the base, including some more *Portalodus* teeth. This was another exciting new discovery as the specimens came from very high up in the Aztec Siltstone section and thus could potentially represent a different faunal assemblage. I eagerly collected anything I could find from this layer and soon my backpack was once more bursting with specimens.

We finished working on the mountain by 6:30 P.M. and returned carefully down the steep slope to our camp. Because of the poor ground definition and constant light snowfalls, the idea of moving camp the next day was slipping away from us. Phil Robbins, our operations manager, told us over the radio that we must consider a twin otter/helicopter pullout option for 10 January. This would involve a twin otter aircraft flying in to drop drums of helo fuel, which then allows the choppers to come in and refuel on the way, pick us up, and refuel again on the way home.

It was still snowing with poor visibility the next morning, so we could not move on. Scott Base informed us that the weather didn't look like changing for at least another day. I was content to spend the day wrapping up fossils, writing up field notes and reading a novel. Later we made up some pasta for lunch, played cards in Margaret and Fraka's tent for a while, and then just lazed around until dinnertime.

At around 8:00 P.M. Margaret made a most unusual dinner of corned beef curry with vegetables and rice. I noted in my diary that after dinner Fraka and I played Frisbee using the plastic toilet seat, and then threw snowballs at Brian and Margaret. We all headed to bed around 10:15 P.M., and I read Brian's book, Robertson Davies' *The Deptford Trilogy*, for a while. The sky appeared to be clearing up, so we were hopeful that we could soon move up the steep pass through the Portal to the Lashly Ranges, a site already renowned for its superb preservation of fossils.

We awoke the next day to clear and sunny conditions with virtually no wind, so immediately after breakfast we packed up camp to make a move to our next site. Leaving Portal late in the morning, we back-tracked over our sledging route until we were a safe distance away from the face of the mountain and its deadly, invisible crevasse fields, then swung around to head directly up the center of the Lashly Glacier. We had a difficult climb up around the front face of the Portal where the topographic map shows the 1600-meter and 1800-meter contour lines next to a small ice cliff. In actual fact it was just a steep incline without any dangers from crevasses, but the incline was too much for the skidoos to pull the two heavy sledges, now laden with many fossil and rock samples. Once again we had to relay the sledges one at a time up the slopes.

That day was a traveling record for us—we covered about 60 kilometers without any problems, not even a single sledge overturn. How bloody nice when things go according to plan for once, I thought that day! As soon as we were well away from the Portal, the sledging became easier on the flat even surface of the Lashly Glacier, which was nicely blanketed with newly fallen snow. From there it was a pleasant day's traveling all the way to Mt. Crean.

We arrived at Mt. Crean close to 6:00 P.M., immediately noticing how much colder it was there. If you look at the map the reason why this is so becomes obvious. This place was on the very edge of the polar plateau, and only the Lashly Ranges were between that vast white elevated surface and us. We were facing the open polar plateau for much of that day's journey, so it became progressively colder as the day wore on.

We pitched camp about two kilometers out from Mt. Crean on the Lashly Glacier. That night the wind started to gust between 20 and 30 knots. This was the katabatic wind rolling down off the polar plateau, picking up speed from the slow gravitational forces that pulled it downwards over the vast expanses of ice.

I thought that a blizzard could be on the way. And indeed it was.

26

At the Crucible of Shark Evolution

> *. . . the inevitable inference was that in this part of the world there had been a remarkable and unique degree of continuity between the life of over three hundred million years ago and that of only thirty million years ago.*
>
> —H.P. Lovecraft

In this passage from *At the Mountains of Madness* Lovecraft suggested that the ancient life of Antarctica may have remained unchanged for millions of years, perhaps unaffected by the several major extinction events that nicely carve up our geological time scale into the neat blocks we call "periods." As absurd as this notion now is in the light of many new discoveries of fossils from Antarctica, which indeed testify that extinction events did occur globally, Antarctica (as the hub of Gondwana) may well have been a crucible for the evolutionary radiation of certain vertebrate groups. In this respect, if the earliest true sharks, called "neoselachians," did originate here, as this chapter suggests, then Lovecraft's fictional hypothesis is not far from reality with respect to this one group. Sharks may well have had an evolutionary explosion in ancient Gondwana, reaching a rapid peak of evolution, then remaining unchanged for many hundreds of millions of years. Sharks only recently disappeared from the seas around Antarctica when the freezing polar conditions set in.

During the VUWAE 15 expedition, Gavin Young, then 26 years of age, collected an extraordinary fossil shark specimen from a small outcrop in the Lashly Ranges, about four kilometers from where we were camped near Mt. Crean. Lashly and Crean were both men on Scott's Terra Nova expedition who, with Lieutenant Evans, made up the return party which headed back to base after Scott and his men sledged on to the South Pole. If not for the bravery and courage of Lashly and Crean, Evans would have surely perished after he developed scurvy and couldn't walk on any further. It is one of the great and often overlooked heroic stories of the early expeditions, well told in Evans' book, *South With Scott*, which he dedicated to Lashly and Crean. At the time this book was published, in the 1920s, it was the only popular written account of that expedition.

That small fossil shark that Gavin Young collected in 1971 became the holotype of a new genus that he described in 1982, naming it *Antarctilamna prisca*. It was also the oldest known shark in the world to have the cartilaginous braincase preserved. The shark was about half a meter long, with one rigid dorsal fin spine. Yet its most endearing feature is its teeth, which are decidedly odd in having three main cusps on each tooth, making it appear closely related to the much younger group known as the "xenacanths." I desperately wanted to find more of this beast or other well-preserved shark remains from the same beds in order to shed some light on the mysterious affinities of this primeval shark. So, naturally, when we approached the Lashly Ranges and saw Mt. Crean peering down at us, I was once more fired up to go searching.

Unfortunately, the "A factor" kicked in again, and a fierce storm descended upon us that night. It eventually confined us to our tents for another three days before we dared to venture out near that sacred mountain of paleontological promise.

We woke up next day to temperatures of 1°F with strong winds gusting at 50 knots. It was way too windy to work outside, so we had to settle down to another long spell of waiting for the weather to clear up. I passed the time writing up some personal notes in my diary. One peculiar thing I noted for this day was that for the last three days while the weather was reasonably good I had been constipated, for the first time on the trip. I suspected that it might have been brought about possibly due to our now higher consumption of dehydrated foodstuffs. Fortunately the affliction ended, except that it was a damn inconvenience because of the blizzard raging outside to have to go and frequently answer calls of nature!

I spent that day reading more on the geology of Mt. Crean and pondering over my notes and map. That night we decided that we would prefer to be pulled out from here on Friday, if given the option, rather than try to go on to Mt. Fleming. It was very cold and blowing hard outside. We recorded gusts of between 50 and 70 knots on our wind meter. I laid awake most of that night because of the incessant flapping of the tent, huddling deep inside my sleeping bags to try and get warm.

The next day saw little change in the weather. It was 1°F with very strong winds when I woke up. We were in the middle of a blizzard. The wind-chill factor outside was now very dangerous, so once more we were tent bound for the whole day. Even very short trips outside for calls of nature resulted in minor frostnip affecting my fingers. I slept in till almost noon, mainly because I'd been disturbed the previous night due to the howling noise of the winds.

I kept thinking how thin the canvas tent was, the sole thing that kept our frail human bodies from the ravages of the blizzard outside. My mind wandered back to a few days ago when we found the remains of a torn polar tent out near Mt. Metschel, and that if our tent suddenly ripped open we would have to make a dash for Margaret and Fraka's tent. The tents are made from very strong material, though, and both held out throughout the night without any damage. As we were the "glamour trip" of the season, Scott Base had let us take the newest tents and the best equipment available for our mission.

All day we read and played cards. After the evening meal I continued reading *The Deptford Trilogy*, and liked the book so much that I made a deal with him to swap it for a copy of my recently published *Dinosaurs of Australia* book. When I arrived back home in Australia I bought a copy from the bookshop, and as I was about to sign it for Brian I noticed that it had the front page in upside down, relative to the cover and the rest of the book. I signed it to him saying that this was the "special edition" for New Zealanders.

I anxiously looked outside next morning. Finally the wind had died down! It was sunny and clear, but still a bit chilly at –1°F. We decided that it was good weather to go out searching for fossils on Mt. Crean. After driving the skidoos across the glacier we pulled up at the base of the huge mountain. Getting up to the outcrops was not as easy as it had looked from back at camp as there were steep cliffs all around the base of the outcrops where we wanted to search. Then we spotted a steep slough of ice and snow which had fallen down one side of the cliffs and which provided an access route onto the section where we wanted to work. Brian led us up to the outcrop by digging footholds in the ice wall with his pick. We put crampons over our boots, roped up and ascended the ice cliff in single file, using our ice picks to stabilize us at each step. Soon we were standing on the flat layers of Aztec Siltstone. We took off our climbing gear and started to fan out over the rocks, keenly searching for fossils.

It was one of the best days on the whole trip. We located the black shale layers found by earlier VUWAE parties and found some excellent fish fossils in them, including the tiny scales of the jawless thelodont fishes, and parts of articulated whole placoderms, rather than just isolated bones. My most exciting find from that level was a weird-looking lump of rock packed full of tiny white scales. Later I was to discover that it was a calcareous nodule full of many hundreds of thelodont scales that could be freed from the rock using weak acetic acid solution. As bone and teeth are made of phosphatic minerals, and limestone is made of calcium carbonate, weak acetic acid solutions will break down the rock but not affect the fossil materials, so after a few weeks of this treatment it is possible to free all the little scales as the rock dissolves so that they can be studied in perfect three-dimensional form. This was the first calcareous layer identified from the Aztec Siltstone to yield such magnificent preservation, not only of thelodont and acanthodian scales, but also shark's teeth, placoderm bones, lungfish tooth plates, and many other kinds of fossil fish remains.

Towards the top of the sequence I found some very well preserved specimens, including a rare articulated specimen of the placoderm *Groenlandaspis* showing the scale-covered tail. Later I discovered in the lab that this layer was also partly calcareous, enabling more bones to be acid-prepared out of the rock, making it another significant locality. By around 6:00 P.M. we had filled our backpacks with specimens and decided to call it a day.

Day 56 began unusually early for us. Brian and I woke up at 4:00 A.M. to listen for the Herc that was supposed to be flying over our area to check out the landing site, but no plane could be seen or heard. The radio schedule at 8:00 A.M. told us that it was late and would be with us in an hour, but it didn't make it over to us at all that day. The weather was fine so we decided to make the most of it and do some more collecting.

We drove the skidoos over to a close-by outcrop where Gavin Young reportedly found the famous articulated specimen of *Antarctilamna prisca*, a section simply called "L1." It is a narrow outcrop of rocks exposed up the otherwise ice-covered southern end of the Lashly Range.

Access to the outcrop was not as easy as it first looked. We had to rope up, attach our crampons and carefully negotiate a pathway through some small crevasses before we had even reached the base of the rocky bluffs. As with the previous day we had to cut ice steps to climb up the steep slopes. After an hour or so we were standing on marvelously flat terraces of light buff-colored sandstone and reddish mudstone that formed a series of wide steps all the way up the hill.

It was another very productive day of fossil collecting. We found two main layers rich in fish fossils. The lowermost layer was very soft, crumbly yellow sandstone that would fall apart in your hands with enough pressure. It was packed full of pure white fish bones, almost perfectly preserved, without any distortion or flattening. I found a few good skulls of the placoderm *Bothriolepis* with many isolated plates of *Groenlandaspis* and randomly scattered scales and bones of lobe-finned fishes. Higher up the section I discovered a gritty sandstone layer rich in small bits of fishes, including many well-preserved sharks' teeth.

One of the teeth I found at this second locality was very peculiar in having a wide broad base with two main cusps separated by a crenulated shearing edge. As I hit the rock to extract the little tooth (only about one centimeter wide) it literally jumped out of the rock into the sandy rubble. I got down on my hands and knees and eventually found the perfect white tooth, completely freed from the rock after 380 million years. Carefully I packed it in a small vial with some tissue paper and kept it safe in my top pocket for the rest of the day. A sketch of that bizarre-looking tooth in my notebook from that day is annotated stating "xenacanth sp. 2 new genus?" Indeed it was. Back in Perth after further study it became the holotype of the new genus, *Aztecodus harmsenae*, meaning "Harmsen's Aztec tooth." By splitting the rocks I found quite a lot of well-preserved sharks' teeth from that layer, and many more when I got the samples back to the lab and could examine them under the microscope.

Already at that time an idea was taking shape in my mind that we had an exceptionally high diversity of fossil sharks present in the Aztec Siltstone. From this expedition we had mustered evidence of at least five different shark genera, maybe more. Previously, rocks older than Late Devonian had only yielded almost microscopic teeth of sharks, usually between one and four millimeters in total size, indicative of small sharks about half a meter or so in length. These older teeth were mostly from marine deposits, but here was evidence that sharks began invading freshwater habitats in Middle Devonian times, possibly for the first time in their history. Some of these ancient Antarctic sharks reached huge sizes for the Devonian, as the largest individual teeth of *Portalodus* suggest it had an estimated maximum size of up to three meters.

The first Devonian fossil shark's tooth found in Antarctica was recovered by geologists Bernie Gunn and Guyon Warren during their 1957 International Geophysical Year sledging expedition to the Skelton Névé region. They were primarily studying the igneous rocks in the region but still managed to bag a few good fossils. One specimen they found was a single tooth from the scree slope of Mt. Feather, which stands across from the Portal, towering 2985 meters high. The tiny tooth was sent to Dr. Errol White at the British Museum of Natural History in London, who established a new genus of shark based on that one bizarre tooth in 1968. He named it *Mcmurdodus featherensis*. The tooth was a jagged array of many flat, sharp cusps along a wide root. Nothing remotely similar to it had been found in any other Paleozoic shark.

However, about 20 years later a similar tooth turned up from the remote deserts of central Australia. Dr. Sue Turner of the Queensland Museum, working with Gavin Young, described it as a new species of the Antarctic genus *Mcmurdodus*. In their paper they put forward a radical idea at the time that these teeth were very advanced in their structure because they possess a hard outer enameloid made of several overlapping layers, a feature that characterizes modern sharks or neoselachians. Up until that find the oldest known neoselachian was a small shark named *Anachronistes* from the early Carboniferous of England, around 345 million years old. The new finds of *Mcmurdodus* pushed the origin of this group back by another 40 million years.

In 1995 I once more made the sacred pilgrimage of all paleontologists to the Natural History Museum in London and searched among the many rows of cabinets full of fossil fishes until I'd located the drawer labeled "*Mcmurdodus.*" I spent some time examining and drawing the little shark's tooth from Mt. Feather. It was only about four millimeters wide and parts of the main cutting edge were broken, but the cusp shapes could be accurately restored with reasonable confidence from what was left.

At the time I was hot on the trail of evidence of any older sharks, and to determine for myself whether the earlier published records of fossil sharks were based on reliable data. The oldest true fossil sharks' teeth at that time were tiny specimens from Spain named *Leonodus*. Nobody doubted these were sharks' teeth, as they had two curved cusps and a well-developed root. Other Early Devonian teeth had been found from Saudi Arabia, and after examining these specimens in the Natural History Museum collections, I was convinced that they also represented several different species of early sharks. However, I also examined some other problematical specimens from the Early Devonian of Canada, first described as belonging to acanthodian fishes (named *Doliodus*), and discovered that they were really sharks' teeth as well.

The interesting link was that the north of Spain and most of Saudi Arabia had actually been part of the northern margin of Gondwana during the Early Devonian. The Canadian specimens may also have belonged to a separate block of crust, called the Alexander Terrane, which was closer to Gondwana at the time than to North America or Europe. The story was starting to jell. Sharks had probably arisen in Gondwana perhaps in the warm tropical seas off its northern coasts, and then they underwent a great radiation and diversification at the start of the Middle Devonian, about 385 million years ago. The first good evidence of this explosion of shark biodiversity was therefore the sharks' teeth fauna found in the Aztec Siltstone of Antarctica and from central Australia. Here was the pointer to the birthplace of all sharks, the very crucible of their evolution.

By the latter part of the Middle Devonian and Late Devonian sharks rapidly became ubiquitous. Although mostly known from isolated teeth, scales, and spines, at least 50 different species of sharks are known by the close of that period some 355 million years ago. From that point on sharks radiated into many different groups, with specialized descendants like the rabbit fishes (holocephalans) and flattened rays. Today we know of approximately 800 living species of chondrichthyans.

My long field season of working in Antarctica's remote wilderness had now come to an end. At around 6:00 P.M. we trudged back slowly to the skidoos, carefully tracing our steps around the crevasses, backpacks laden once more with the heavy spoils of our hunt. We had found many new species of fossil fishes, fossils which would now be put to good use back in Australia by contributing to solving problems of global plate tectonics, past climates, and age correlations right across the broad realm of Gondwanan countries. Inwardly, I felt quite pleased that the whole trip had gone well scientifically, but as we headed off back to our little yellow pyramid tents in the distance, I knew that our work was not yet over until our specimens were safely back at Scott Base.

That evening our 8:00 P.M. radio schedule informed us that VXE-6 was planning a recon trip over to us the next morning. If they were not happy with the potential for landing a Herc on the ice we would have to try to get two New Zealand Air Force helicopters to pull us out. At this stage, because an airlift was so crucial to getting all of our specimens and equipment out in one go, we were told to report back with weather conditions to Scott Base every six hours from then onwards.

That night Brian and I joked about how we could lure a Herc down to pick us up. We concocted a somewhat silly advertisement that could be broadcast to any approaching plane. We were both delirious over it, laughing uncontrollably, although this may have been due to other factors acting on us at the time. So, anyhow, I wrote the whole inane conversation down in my notebook. In the interest of future psychiatry students studying the mental health of deep field expeditioners, here it is:

> Hey guys, it's Bingo Night down here at the Lashly Glacier—Free Parking—Don't forget about our special this week—a Free Lunch—succulent turkey, a big super deluxe block of chocolate, barley sugar, and hey, guess what? We've just had ourselves a bit of a whip around! Yessiree, we've got a big hat full of greenbacks—just a few bucks for you guys that work soooo hard for VXE-6.
>
> Now, why don't you come down here—right now! This glacier's smoother than a baby's arse—why you could land a C5 here no worries!
>
> It's Pick-Up and Pullout time for K221 and Drop-Down and Collect Time for XRDs!
>
> Hey! It's Hootenanny Night down here on the Lashly Glacier! Going to McMurdo? We're pretty friendly down here—we've got some good lookers that want to go all the waaay!

As the diatribe shows we were really quite silly by this stage and even discussed in detail the merits of the C-130 Cargo Cult, suggesting that we needed to build a full-scale C-130 out of snow and place it down at the start of the ice runway to appease the fussy gods of VXE-6. Needless to say, I slept very well that night, still chuckling to myself from deep within the confines of my two sleeping bags. Brian awoke every six hours and radioed back details of the weather.

Our last full day in the field was Friday, 10 January 1992. The weather was fine, clear and sunny with almost no wind, but a brisk –1°F. Visibility was excellent, we could see for about 30 kilometers, and there were only a few wispy cirrus clouds loafing around a very bright blue sky. The plan for the day was to pack up and prepare for the pick-up, plus to make regular weather reports to base. At 10:00 A.M. the temperature plummeted to –6°F. The Herc was due to arrive at 11:00 A.M. but was delayed. We decided to pack up our base camp and move out to the middle of the Lashly Glacier to facilitate the pick-up.

That afternoon Brian and Fraka drove the skidoos up and down the middle of the glacier for about an hour to flatten out a runway area for the Herc. Anything at all that we could do to entice them down to an easy, safe landing was worth a try, according to both Brian and Margaret's past experiences of many delayed pick-ups. The runway they made that day was about ten meters wide by a kilometer long and looked so good that you'd have thought a jumbo jet could land safely on it. How deceptively perfect it looked! Little did we realize then the problems that runway would soon cause us.

Herc XRD-04 came droning overhead about 4:40 P.M. It did a quick fly-past, and the crew chatted to us on the radio for a short while. They seemed to be very happy with our landing strip so planned to return early the next morning for our pick-up at around 8:00 A.M., provided the weather was fine.

From here on we were instructed to give four-hourly weather reports back to Scott Base, followed by hourly reports throughout the night. That evening Fraka cooked what we hoped would be our last supper in the field, a dehydrated beef curry with soup, peas, and other dried vegetables. We were all very excited by the prospect of going back to Scott Base and the luxuries of civilization, but were inwardly doubtful of actually being picked up on time. Usually there are delays for at least one or two weeks, as exemplified by Margaret's experience on the Darwin Glacier in the 1988-89 season. However, if the good weather would hold out for one more day we just might actually get pulled out on time. We never gave up hope of something actually happening according to schedule, as for example our navigating the Mulock Glacier, which went ahead right on time.

The fine weather continued that night, but it also had its downside, as it became uncomfortably cold. I recorded our coldest temperature for the whole trip on that very last night. At 3:00 A.M. I popped my head and upper body out of the tent and spun the thermometer around for a full minute, as was the routine practice. The sky was still quite clear with about three-eighths cloud cover and the air temperature was –18°F. I grabbed a few hours of intermittent sleep, occasionally waking to hear Brian's hourly weather reports.

On the morning of 11 January, at 5:00 A.M., it had "warmed up" to –9°F. The sky was amazingly clear, not a cloud to be seen anywhere, and no wind. Sunshine was belting down on us. It was *perfecto mundo* for a Herc landing!

I was full of hope that we would be going back to Scott Base that day.

27
Pick-Up Day Problems

My whole body is apparently rotting from want of proper nourishment—frost-bitten fingertips festering, mucous membranes of the nose gone, saliva glands of the mouth refusing duty, skin coming off whole body. The sun bath today will set much right however—I felt the good of the sun as I have never done before.

—Douglas Mawson

Seventy-nine years earlier on that same day, 11 January, a young Douglas Mawson was in a pretty bad way physically, yet some how he managed, against all odds, to keep on man-hauling his sledge for another three weeks before finally stumbling back alive to the hut at Cape Denison on 8 February. The appreciation of warmth and comfort of the sun was something we all shared with Mawson, along with the one preoccupying thought that day of heading back home to the base.

We started the day in good spirits in anticipation of going back to Scott Base without any hitches. Everything had been packed up except for one tent and we were eagerly waiting for the radio report that the Herc was on its way. It was a tense time as although the skies were clear around us, there could well be a howling blizzard back at McMurdo and we wouldn't know about it. Suddenly at 9:20 A.M. the radio crackled into life. It was Hercules XD-01 giving us only 20 minutes' warning for the pullout.

We quickly sprang into action and packed up the last tent, tied it onto the sledge and moved everything out near the runway. We heard the plane's low grumble overhead only minutes later. It landed easily on our nicely cleared ice runway. We had packed all our gear on the sledges with the idea of just running them straight up into the back of the Herc as we did on put-in day. This part of the operation went smoothly. After about fifteen minutes all the sledges and skidoos were safely on board.

We climbed inside the Herc, sat down with our seatbelts on and waited eagerly for the take-off. I was stunned that everything had actually gone according to plan, as scheduled, without a single hitch. Margaret, Fraka, and Brian were also very surprised by this, as they had expected the customary delays due to unforeseen weather changes, or last minute mechanical repairs to be done to the planes. It seemed too good to be true.

The Herc then powered up to full speed and tried to take off, but without any success. We had calculated that we were carrying only about half the full payload of permitted weight for the Herc, so this shouldn't have been the problem. Once more, the plane turned around and tried to power off the ice runway. No go. After a few more unsuccessful runs, the pilot next tried to bounce the plane a little to settle down the soft snow which was causing the drag on the plane's skis. We suddenly felt the plane bottoming out as it bounced up and down on the snow, jarring our gear and us in the cargo area. It seemed like our "perfect runway" was fine for little skidoos, but it had a thick blanket of newly fallen snow underneath it which was causing great frictional drag on the plane's huge landing skis.

The crew then asked us if we would mind moving into the tailgate of the plane. They politely beckoned us to sit over the large rear-opening door of the Herc, and then strapped us down flat with a large cargo net. Now I know exactly what it's like to be a tea chest. The captain thought this would place our weight more rearward to give us that little bit of extra nose lift required. Once again we tried to lift off with engines roaring at full power, but no cigar.

The plane then filled with "smoke" (so I thought) and, as we were strapped down with a net over us, we watched with growing concern as the crewmen put on their oxygen masks. I imagined the plane was going to go up in flames or something, but apparently it was just that the intake vents of the engines had taken in some snow, causing a mixture of smoke and water vapor to fill the cabin!

After one-and-a-half hours of unsuccessful attempts to take off, the captain gave us the sad news that he would have to dump all of our equipment and specimens outside in order to take off because they were now getting critically short of fuel. This was disturbing news as we had recently heard the story over the radio of Paul Fitzgerald's party (event K5076) in northern Victoria Land, who were forced to leave all their gear and samples behind after six grueling weeks of fieldwork. They never did get their samples back as it was deemed too dangerous to try and land a Herc in that region again.

At about 1:15 P.M. that day we arrived safely back at Scott Base, but instead of the elation of returning home triumphant with our specimens, we were all deeply concerned over the fact that all our precious geological specimens were still out there in the middle of the Lashly Glacier. It felt strange for us to be back at Scott Base; a bitter-sweet melancholy descended over us. It should have felt great to be back, but we were sad because unless our specimens were retrieved the whole three months we had just spent down in Antarctica would be wasted!

I first noticed how hot it was inside the base. Then I noticed that there were lots of people around. It felt very different from our simple life in the field. I wrote in my notebook that it seemed "hard to adjust" to being back in "civilization" again.

The first thing we did was to take a long shower, our first in over two months, and change into clean clothes. We refrained from drinking at the bar until news came from VXE-6 Squadron about our gear. Then the miracle we had hoped for happened. At 8:00 P.M. we received a phone call reporting that VXE-6 had made a special effort in view of the incident in northern Victoria Land and had gone back to the Lashly Glacier to successfully pick up all of our gear. It was now waiting for us over on the ice runway at Willy Field! We were immensely overjoyed. It meant so much to us that our specimens and equipment were finally back safe. Immediately we hitched a ride over to the airfield. It only took us about fifteen minutes to unload our skidoos and sledges from the Herc, then we drove our sledge trains elatedly back to Scott Base.

A lot of people were in the bar watching our sledge trains come around over the sea ice as we pulled up at the field store hangar. Some of the important items of equipment, such as the radios and first aid kit, were unloaded and placed safely inside the shed, but the majority of the gear was left on the sledges. There would be plenty of time to sort it all out the next day. Eagerly we adjourned to the Scott Base bar around 9:00 P.M. for a few celebratory drinks. We were so happy, words could not do our feelings justice.

We stayed drinking and chatting till the bar shut, then kept on reveling with some of the others using our own grog supplies. We enthusiastically recounted all of our adventures, right down to the last details, to anyone who wanted to listen. Eventually, dog-tired and feeling somewhat mentally and emotionally frazzled, I crawled off to bed around five o'clock in the morning.

28

Base Blues and Arrival Home in Australia

It is now my terrible duty to amplify this account by filling in the merciful blanks with what we really saw in the hidden transmontane world—hints of the revelations that have finally driven Danforth to a nervous collapse.

—H.P. Lovecraft

Madness, so they say, is just a relative degree of sanity. After a long spell out in the wilds of Antarctica, one's sanity can be strained from the pressure of constant danger, the long hours of working and from the endless days of boredom spent tent bound when it is bad. Although none of us went "mad" in the traditional sense, it would be true to say that my emotions and feelings were definitely heightened, sitting far above their normal background levels.

After a day of rest we spent Monday dutifully unpacking our specimens from their boxes and checking that everything was properly labeled with its correct field number. All our equipment had to be returned to the supply shed. This involved reporting any damage, doing minor repair jobs, and checking everything off the equipment lists with the store man. We finished this job by about lunchtime.

The total weight of the specimens I had collected for the Western Australian Museum came to nearly 380 kilos. After rewrapping each of my fossils with newspaper and bubble plastic, I carefully packed them into five large wooden crates, screwed down the lids and put my museum's Perth address on each box. The icebreaker ship, transported first back to Christchurch, and then freighted on to Western Australia, would soon pick these up. They eventually arrived in Perth about three months later with every specimen in good order.

That night, in preparation for departure, we took our bags to Willy Field to weigh them in for the flight, whenever that would be. A late flight back to Christchurch was scheduled that night with us listed as "low priority" passengers. After the airport formalities we returned to Scott Base to uphold the long-standing tradition of ringing the bell and shouting the bar. We stayed up waiting for our call to the airport that never came. So, we kept on partying till well into the wee hours of the morning.

All four of us were feeling the dire effects of homesickness. It was disheartening knowing that our work was completed but that we had to just wait around until placed on a flight home. My state of mind was still on a high from the trip, and perhaps we all enjoyed the comforts of base a little too much. I went to bed disgruntled at about 6:00 A.M. that morning after staying up all night talking and drinking.

The next day we were informed that we were going home that night. Because our bags were still at the airfield each of us had only a limited range of clothing so were prevented from doing the many outdoor fun things that the Scott Base environs has to offer, such as cross-country skiing, long walks on the sea ice, or skidoo rides.

The officer in charge assigned us various base duties to keep our minds off the eternal thought of going home. My first task was to write an article for the Scott Base newsletter on the fossil fishes of Antarctica, which I did in the science lab on their Mac computer. I called the article "Chips of Fishes" and it was published in the following issue of the *Scott Base Times.* It was absorbing for a short while to bury myself in this task and combine all my thoughts about the fossil fishes in one simple article. Using a computer in Antarctica, even inside the base, meant having to wear a wristband to earth oneself from ever-present static electricity shocks.

We were told later that our scheduled flight home had been cancelled and that we may not get onto another flight for two more days! Somehow we were expecting this. That evening we were all feeling very despondent about the whole situation. We were well and truly experiencing the downside of the long high from the field work. I went to bed very early, about 9:30 P.M., but didn't sleep very well as it felt uncomfortably warm and stuffy on the base. I suppose our bodies had physiologically adjusted to being in field conditions and were then slowly readjusting to normal again.

During those few days that followed we kept feeling flat. Hanging around the base waiting to go home was excruciatingly monotonous, especially after the adventures, discoveries, and excitement of two months' working by ourselves out in the remote deep field. This passage from Nansen's *Farthest North*, taken from his time during the long, boring winter months, expresses some of the feelings of anguish that we were going through at the time:

> I know this all a morbud (*sic*) mood; but still this inactive lifeless monotony, without any change wring's (*sic*) one's very soul. No struggle, no possibility of struggle! All is so still and dead, so stiff and shrunken, under the mantle of ice. Ah! . . . the very soul freezes. What would I not give for a single day of struggle—for even a moment of danger!

On Wednesday, 15 January, we were assigned a highly dangerous mission, one that only people of our high level of academic training and field skills could dare to attempt. Our orders came straight from the top. We had to clean all the outside windows of Scott Base. It was a monotonous job, but I suppose it served the purpose of keeping our minds temporarily off the burning thought of going home. I recall it being a relatively fine day, the late summer sun shining brightly and hardly a puff of wind about, so it actually felt good to be outdoors doing something physical. My mind often reflects on my life as a scientist, combing the far reaches of my hazy memory for all the odd and crazy things I have done in the name of science, like that day when I was totally absorbed cleaning windows.

After lunch we received the welcome news that we would be going home that night. Then, the biggest let-down of my life came only moments later when they added that only Margaret and Brian were scheduled to go on that flight. I felt thoroughly miserable again. Margaret then had some harsh words with the boss there, Dave Geddes (the in famous man who gave me the piss bottle on 7 December), and Dave somehow managed to pull some strings at McMurdo to get the manifest changed so that both Fraka and I were put back on the departure list for that night. However, despite this good news, there is always some catch! It seemed that our bags would not be going with us. The thought then arose that we would get back to New Zealand, but I would probably have to wait around for several days until my gear arrived, an even more depressing thought when all I wanted was to get back home to Australia and be reunited with my family as quickly as possible.

Priority seating for flights home was always such a hit and miss thing. Anyhow, we were deliriously happy to be scheduled on a flight home once more. Later that evening, we were overjoyed to hear the news that not only were we all confirmed to go on the same flight, but that our baggage would come with us after all!

At about 8:00 P.M. that night we said our last goodbyes to the folks at Scott Base, then drove down the icy road to Willy Field. After several hours' waiting, we finally boarded the Hercules aircraft around 11:30 P.M. for the midnight red-eye special to Christchurch. Once more strapped in like sardines for take-off, we were more than anxious to get that long flight back to civilization behind us. There were only a dozen passengers on the flight that night because it was mostly carrying cargo. After take-off we all spread out to find a comfortable spot to sleep.

I don't remember anything about that flight home, apart from the fact that it took about nine hours. I know I didn't sleep much because I'd left my move too late to grab a comfy spot, so I was sort of hunched over on my seat against my bag. I didn't talk to any of our field party because the noise of the engines was too loud, so I just brooded in my thoughts, catching snatches of broken sleep whenever the tiredness overtook the discomfort of the cramped seating.

The plane touched down on the tarmac in Christchurch at around eight o'clock the next morning. Still dressed in our Antarctic clothing, we were taken back to the DSIR headquarters to change and return our field clothing. I immediately rang Qantas to get myself placed on the next flight back to Sydney, which wasn't until 6:00 P.M. I said my goodbyes to Margaret, Fraka, and Brian; each of us exchanged powerful hugs. Few words were said. We'd shared wonderful experiences that were now set in stone deep within our minds. I felt that we were all strongly bonded by our experiences, that none of us would ever forget each other's friendship from that long expedition.

Everyone went their own way, eager to catch the first available flight home, so I was soon left alone to bide time at the international airport most of the afternoon. I had plenty of time to pick up the obligatory bottle of duty-free grog, buy myself a good book to read and just relax, doing nothing much for the first time in three months.

By 6:00 P.M. the plane was ready to leave for Australia. As we rolled down the runway pointed for good ol' Australia, I felt powerful emotions welling up inside me, churning my stomach and bringing uncontrolled tears to my eyes. This was to be a pattern of unexplained emotional behavior that I had almost no power over for the next few months, and which even now can surface at odd times, like when I'm watching an emotionally powerful movie, or just having strong thoughts about my kids or someone I love. I suppose Antarctica did this to me by raising my inner feelings to heightened levels they'd never before experienced. By facing such extreme challenges, my emotions had been tested to the limit when I'd faced the possibility of death on several occasions, and had hastily pushed these experiences to the back of my mind so as to get on with the job of survival. Eventually they had to surface again, and when they did it was in ways beyond my control.

I arrived in Sydney and was warmly greeted by my brother-in-law, Tony, who put me up for the night in his Drummoyne apartment. It was a groovy apartment with a superb view of the Sydney Harbour Bridge, which spanned across the first night sky I'd seen in three months, lit up like a Christmas tree. Over a few beers that night I recalled the highlights of the whole trip to him, the first of many such retellings, which after awhile would become almost rote as people asked me the standard question: "How was Antarctica?" They still do, so writing this book is one way I'm dealing with it.

I discovered to my delight that Tony, who worked for Qantas, was going to be my pilot for the flight back home to Perth the next morning. Even better, he managed to have me upgraded to business class for the final leg of my trip home. Just before take-off I was invited into the cockpit and shown to the jump seat where I could indulge my taste for excitement one last time by watching Tony, then one of Qantas' youngest first officers, take the gargantuan plane up into the sky. Later on arrival I was also able to see him land the great thing safely on the tarmac in Perth—a spectacular way to end a grand adventure.

When I stepped off the plane at Perth airport and saw my wife and kids for the first time in three months, my eight-year-old daughter Sarah immediately burst into uncontrolled laughter, followed by my other kids and my wife. They had never seen me with a beard before, and probably never will again, as it looked really bad. I appeared to them like some scraggy hermit from the mountains of Tibet. My hair was longer than it had ever been, hanging down around my shoulders, and my face sported a weird goatee-like wisp of a beard and a fair moustache. Not a good look.

I was emotionally overcome to see them and after hugs and kisses and tears shed all round, we went back home.

Although Antarctica was now behind me, I was never beyond Antarctica. Not ever again. My life changed from that moment onwards.

29
So Much for the Afterglow

Perhaps we were mad—for have I not said those horrible peaks were mountains of madness? But I think I can detect something of the same spirit—albeit in a less extreme form—in the men who stalk deadly beasts through African jungles to photograph them or study their habits. Half paralyzed with terror though we were, there was nevertheless fanned within us a blazing flame of awe and curiosity which triumphed in the end.
—H.P. Lovecraft

Lovecraft writes in this passage about how the spirit of investigation, the flames of curiosity, eventually triumphed over the fear and adversity of their ill-fated expedition. I can empathize with his words, as the passion to research our new finds would eventually triumph over the lingering adverse effects of our expedition. However, straight after my journey through the mountains of madness, my first task was to settle back to normal life once again.

Despite the contrast of being thrust into the middle of a blazing Perth summer from the cold depths of Antarctica, it felt truly great to be home; to live with my wife and kids every day. Yet it was probably a few months or so before I fully readjusted and could once more feel comfortable with daily routine. After all, for three months while on a deep field expedition you don't have to worry about money, paying bills, the house, the garden, your family, or even where your next meal is coming from. Your mind becomes attuned to just the basic needs of meeting each day's challenges in order to achieve your results.

Physiologically your body adjusts to the extreme cold after a few weeks out in the field, soon becoming used to a high fat, high kilojoule diet. Despite eating huge amounts of food every day that I was in Antarctica, I still lost nearly a stone (about 14 lbs) in weight due to the sheer output of energy from the work and from my body constantly burning fuel to keep warm. Every breath of chilled air has to be warmed by your body, so just breathing in Antarctica consumes a lot more energy than in warmer climates. To physically and mentally adjust once more to one's normal climatic and social conditions is actually harder than you would imagine.

Back home I found myself always talking about Antarctica to my family and friends. Eventually I had to be made conscious of the fact so as not to bore everyone to death all the time with "this reminds me of when I was in Antarctica . . ." or "we did it like this in the Antarctic . . ."

There's no denying it is an amazing place. After having wild adventures and being on a high from collecting so many intriguing specimens, it is hard not to view regular life as anything but mundane. Many seasoned Antarctic expeditioners say that they won't be going back down there ever again, but next season they are the ones first in line for the plane. It is something hard to explain, but the old cliché, "it gets in your blood" will suffice.

For about three months after returning to Australia I was on a bit of an emotional roller coaster, sometimes crying without reason at sad movies or emotional events, or laughing wildly at simple, stupid things. Maybe my mind was just letting off the accumulated steam of the trip. My emotions eventually seemed to have mellowed out into the realm of normality but, to be honest, have never really been exactly the same as before the trip.

One idea that started cementing into my mind at that time was the concept of our body's vulnerability. A few weeks after my return to Perth I attended the funeral of Dr. Nicholas Rock, who had committed suicide in early 1992. I had known Nick reasonably well when he was a lecturer in the Geology Department of the University of Western Australia, where I was a research fellow during 1986-87. Together we had publicly debated the creationists. Anyhow, I remember being in quite an emotional state at the funeral because I kept imagining all through it that it could well have been my funeral, and that a split second's hesitation out in the crevasse field had almost been my undoing. Furthermore, had that scenario really happened, my body would probably not have been able to be retrieved.

These feelings also hit me hard again in late 1995 when I was working in Paris for three months and preparing for a quick trip to Dublin to visit one of my old friends, David Johnston. Dave and I were at Monash University together doing our doctorates in the early 1980s. We shared houses and became very close friends who stayed in close contact over the intervening years. After completing his PhD thesis Dave was appointed as a lecturer in structural geology at Trinity College. He had arranged for me to come over and give a lecture to the Geological Society of Ireland in mid-October. About ten days before I was due to go to Ireland I was busily working away in the collections of the Natural History Museum in London when there was a phone call for me. It was the news that Dave had disappeared off the west coast of Ireland while carrying out geological research work. They found his car, his dog, his notebook and a few other things, but no Dave. To this day no trace of him has ever been found. It was presumed that a freak wave must have swept him from the cliffs into the cold North Atlantic. This devastated me at the time, as Dave was about my age and had been one of the most dynamic, fun-loving, and intelligent people you could ever meet.

Although shattered, I did venture over to Ireland and gave the lecture in honor of his memory. A friend and I hired a car and drove over to the west coast, where I stood upon the jagged rocky cliffs of Annagh Head, and said my goodbyes to an old friend. Once again, life's vulnerability came crashing down upon me and filled me with gloom. Today I have overcome these melancholy feelings largely through the passage of time, my involvement with my children and a renewed interest in the martial arts. The concluding words of Admiral Byrd's book, *Alone*, after his near disastrous solo stay over winter in a small hut in Antarctica in 1934, have a deep ring of truth to my mind: "So I say in conclusion: A man doesn't begin to attain wisdom until he recognizes that he is no longer indispensable."

Antarctica still features prominently in my work today. The Western Australian Museum houses a large collection of fossil specimens resulting from our expedition, many of which await detailed study. Because the specimens were difficult to acquire I regard this is a job worth doing well, without the regular pressure of having to rush to get it all published. Over the last few years I have published a number of scientific papers on the new material, and more work is in progress. I am also working on a book documenting the complete history of life in Antarctica,* and have been carrying out further research on how the fossil faunas from Australia and other Gondwana countries, like South Africa, and Iran, relate to those from Antarctica.

On return to Perth in 1992 I put together a small exhibition on Antarctica at the Western Australian Museum. This comprised a collection of my best fossil fishes, rocks, and mineral samples, various items of Antarctic ANARE-issue clothing and some of my best photographs enlarged and block mounted. We augmented the display with other specimens we already had in our collections, such as an Antarctic meteorite, some fossil sea urchins from Seymour Island, and a stuffed emperor penguin collected on Mawson's 1910-14 expedition. Curiously, the most popular item in the display was a set of my underwear, stretched out with pins on the wall and boldly labeled as "John Long's Long Johns."

My memories of Antarctica still serve me well in daily life. Every time things seem to be getting difficult, like the mounting stress of work, too many commitments, large bills to be paid, and so on, I realize how it is all just a drop in the ocean compared with life's big picture of working in a harsh environment where one's life is at risk almost on a daily basis. I still thank my lucky stars each day that I'm here to enjoy spending time with my children and am able to watch them grow up.

If anything, I've possibly become more ascetic through my experiences in Antarctica. For the last four years I have lived in a small apartment in inner city Perth that has a marvelous view over a scenic park adorned with many trees and a lake. My living space is small and my possessions are few. I have often pondered whether the resemblance between my living conditions in a polar tent in Antarctica and my present arrangements in modern society is mere coincidence.

---

*For Johns Hopkins University Press.

In 1993, just twelve months after I came back from Antarctica, my marriage of eleven years began to fall apart and abruptly ended that year. It was an extremely difficult time in my life and it probably took some years before I could honestly say that my life was in good order again. These days I see my three kids often, both at weekends and during the week when they stay with me, and I know that now they are happy and well adjusted to the major changes that took place back then. They are all doing very well at school and I often bring them along into the field with me when I go searching for fossils.

I don't think my life would have fallen into place again so well without the experiences and inspiration that I gained from my time in Antarctica. After coming back fitter and healthier than I'd been for a long time, I decided to try and keep myself in good shape, so I took up karate, a sport I'd enjoyed many years earlier as a teenager. During the difficult years of my separation and divorce my *sensei* (teacher), George Chaplin, worked me much harder than usual, turning the physical training into my emotional therapy; changing the build-up of negative energy one has at such times into positive, muscular strength. I am now a black belt (3rd *dan*) and have become the chief instructor for our style, Uechi Ryu Karate, in Western Australia. I derive great satisfaction from training and teaching classes four times a week to people of all ages, including my own children. It is especially gratifying to see my students grow stronger and become more self-confident. In August 1999 I spent three weeks in Okinawa, the birthplace of karate, to train with the venerable masters of my style, and participated in an international karate tournament as a member of a small Australian team. It was a highlight in my life that epitomized my current philosophy of following my dreams when I can, rather than going along with the normal expectations of society, such as having to own a big house, a new car, or the latest computer.

Karate has taught me more than just being able to keep my body fit and flexible as I enter my fifth decade of life. It has taught me how to sharpen my mind. I had for many years believed the concept of "mind driving the body." This hypothesis purports that if you have a sharp mind and can pursue your intellectual studies to fulfillment, your body will follow. Not so. I now believe the opposite hypothesis is more important to achieving my long-term goals. A strong, healthy body pushes the mind to higher degrees of functioning because the nervous system is in better condition, which can heighten one's powers of thinking and reasoning.

I had never realized this until achieving a high level of fitness through years of regular karate training. Now I have the energy and enthusiasm to pursue my paleontological research, my karate, and my writing while balancing these interests with time for my children, family, and social life. Doing something now, rather than later, is also foremost in my life, as is the pursuit of the physical and intellectual goals in life above the material things. I hold that it's not what you own or whom you know that is important, but rather what you do and how it positively affects other people's lives that counts for life's bigger picture.

My simple philosophy of life these days is that our personalities are made up of four corners. The four corners are the intellectual, physical, emotional, and spiritual components. The "physical corner" is simply keeping physically fit and healthy. The "intellectual corner" is keeping your mind active through regularly pursuing problem-solving tasks. The "emotional corner" encompasses your relationships with your family and other people. The "spiritual corner" is simply one's ability to cope with the unfathomable things of life, either through religion or logic. To balance them in the correct ratio is the goal of every individual, and this ratio is naturally quite different for every person, as no two of us on this Earth are the same. One's harmony of life will be achieved when the four corners merge to form a perfect circle, all components being in the correct balance for your life's particular direction. I have no idea how one will know when this is achieved, except that I expect you will feel that you are happy with your lot.

This concept in itself is an enigma, but then again, so is all of life. Living is never straightforward or simple. We should not expect to completely understand or try and predict the directions our lives could go in. One can only ride with the bad times, be thankful for the good times, and keep on existing in physical terms as comfortably as possible.

I think everyone has their "Antarctica," whether it is the real place they have visited, a challenging experience in their lives, or an uplifting feeling in their minds and hearts.

30
Reflections from the Ice

The world has only two frontiers left: the seabed and Antarctica.
—K. Suter

Four billion years of evolution have culminated in where we humans are now, although as a modern species we date back only about 120,000 years or so. Humans have inhabited all the major continents, except Antarctica, for many thousands of years. Antarctica has been home to our species for less than a century, but in the 1990s less than a thousand people a year actually inhabit Antarctica for a year or longer.

Perhaps, though, one day in the distant future, if unfortunately our ecosystems have been destroyed by too much clearing of forest, loss of irreplaceable soils, increased salinity in ground waters, and the excessive pollution of the atmosphere, the human race may look at Antarctica with hope. It could be possible for people to inhabit the vast wilds of Antarctica in large cities built on the polar plateau, harvesting the continuous months of summer sunshine as energy and growing food in vast hydroponic biospheres. Wind energy and solar power could be harnessed to heat the dwellings and provide enough energy for daily needs. In fact there is so much energy in Antarctica from its fierce katabatic winds and five months of almost perpetual sunshine that maybe one day, when solar and wind energy technology is affordably cheaper to transport than it is today, Antarctic energy could be channeled back to other parts of the world.

In this way Antarctica could potentially provide the human species with some respite while the rest of the planet commences its long period of recuperation from the decades of ecological degradation. This, of course, is my worst-case scenario, and I sincerely hope that the earth's ecosystem doesn't get into this much trouble over the next few centuries!

Let us at least entertain the concept of Antarctica as a possible last resort for human habitation. It would be far cheaper to live on the polar plateau of Antarctica than to live under the sea in pressurized cities or to colonies the moon. It is certainly large enough to hold billions of people, once we have the technology to live there self-sufficiently, presumably well away from the sensitive near-shore and shelf ecosystems that most of the wildlife inhabits. It has a vast source of freshwater and sunshine that together, with input from fertilizers, can provide the basis for a hydroponic agriculture. Monitored harvesting of food resources around the continent combined with human and animal waste could provide protein and fertilizer for this system.

There is no reason why the vertebrate life of Antarctica could not one day be commercially farmed as food resources, once breeding populations are high enough. Whales could then become the cattle of the southern continent, seals the sheep, and penguins the poultry. I know these ideas might appear shocking to some ecologists and environmentalists, but they are just mere thoughts, not suggestions or anything that I could imagine happening in the near future. It is far more ecologically sound to farm the animals that are native to the Antarctic ecosystem rather than introduce foreign species, such as polar mammals from the Northern Hemisphere. The point should at least be made that we must never overlook Antarctica's huge potential as a possible last resort for humanity's survival.

Antarctica's vast biomass is indeed a potential food source to feed millions of people, especially if they were based on the continent itself. The rich shelf seas around Antarctica are home to an estimated 5,000 million tons of krill, *Euphausia superba*, a small shrimp-like crustacean which forms the base of the food chain for much larger creatures like the baleen whales, but also for seals, sea-birds, fishes, and other invertebrates like squid. Krill has been commercially fished for human consumption by the Soviets, who in 1981-82 took around 500,000 tons. Although krill can be very high in fluoride, so is not directly safe for human consumption, its value might be realized more in processed form as a cheap source of protein for undeveloped countries, if costs of procuring the harvest and processing it can ever make the final product safe and affordable to such countries. Alternatively, it could be harvested in the southern oceans to feed the whales to feed the humans. . . .

More recently studies by scientists at the British Antarctic Survey's Marine Life Sciences Division have demonstrated that the relative abundance of krill from year to year directly reflects the scale of bioproductivity of the Antarctic marine ecosystem. In years of low krill abundance the whole food chain is affected—there is a lower abundance of certain fishes, whales, seals, and penguins that rely on the krill as their main food source. Such events can sometimes have an effect on us humans, as some of these species, like the Antarctic cod, were commercially fished back in the 1960s for our consumption. So, the study of krill will no doubt be very important for human food resources in one way or another.

In 1980 all signatories to the Antarctic Treaty signed the convention on the Conservation of Antarctic Living Marine Resources and it was ratified in April 1982, largely prompted by the heavy Soviet krill fishing. The living resources covered by this document include all things except whales and seals, which are both covered by different conservation conventions. If krill fishing was ever firmly established in Antarctica, there is the strong possibility that having hundreds of large ships continuously in dangerous southern waters could result in some major disaster, which could lead to the pollution of these waters. Despite this, the rest of the Antarctic food chain has not been thoroughly enough investigated as to its food value for human consumption. Obviously there are many millions of tons of fish and squid, not to mention other consumable creatures, but we need more research in respect of resource estimates and sustainability for long-term harvesting. Nonetheless, Antarctica's food resources are a major factor to be considered in all future planning about how to feed a growing human population; more so if humans ever populated Antarctica in a major way.

The development of tourism to Antarctica is of growing concern to some people, but I feel this is a necessary process to allow more people access to its magnificent grandeur. If tourism is controlled by confining most visitors to ships that retain all their waste products and by carefully monitoring the presence of visitors when on short forays to the mainland, then I see no real environmental problems arising. Some may argue that just the presence of large commercial ships in Antarctic waters is a potential threat due to possible oil or fuel spillages, and with this I cannot disagree, but I think that in an age of increasing technology it could be done with little risk to the environment. Perhaps, in the not too distant future, large sailing ships boosted by solar-powered engines could safely take tourists to and from Antarctica.

The number of people who currently visit Antarctica as tourists is minuscule by comparison with the number of tourists to other parts of the world's densely populated regions, such as the Mediterranean or the small Pacific islands of Fiji, Tahiti, and Hawaii. Cost more than anything currently keeps in check the number of people who visit Antarctica as tourists; and, by virtue of its extreme climate and distant southerly location, cost will continue to be a major factor limiting the numbers of visitors for some time to come.

Personally, I believe that Antarctica should not be the domain of the privileged few, but be accessible to the people of the world just as any other great wilderness, so long as tourism is continually monitored to prevent any adverse impact on its ecosystems.

Antarctica, for me, is a metaphor for the unblemished corner of humanity's soul. Its conquest has cost many lives, but in that process we have redefined the very nature of heroism, bravery, and discovery in a new context.

We need Antarctica. We need to preserve it.

In all its pristine, white dignity.

Epilogue

Teardrops, emotion-tainted water
Now ice within ancient rocks
Mantled by unblemished snow
Fear, hope, love, and ice
Have crystallized as one
The body fades into non-existence
Yet traces persist
Forensic memories
Life, eternally,
Frozen in time

On a flight across Australia I kept hearing the song "Teardrop" by the band Massive Attack on the plane's popular music channel. I'd seen the band in Perth earlier that year and had bought its album, but I'd never really listened to the actual lyrics of the song. As it was a long flight I eventually heard the song about four times, and then the words finally sunk in. The line "teardrop on the fire" set me thinking about my experiences of Antarctica, and inspired the poem above.

On 1 January 1992, the day on which I survived falling into a crevasse and being almost buried in an avalanche, my tears had fallen onto the rocks and snow. They would have seeped into the microscopic pores of the 380 million-year-old sandstones and would still be frozen there today, protected within those minute cavities, even if covered and somewhat ablated by the snow and ice from the ravages of several harsh winters.

The realization then dawned on me that a small part of my essence, of my molecular body and my ethereal soul, will now remain forever in Antarctica.

Appendixes

Appendix 1
What I Liked and Disliked About Antarctica

This list was written down in my field notebook after returning from the major sledging journey of 1992 while waiting around on Scott Base.

Things I like about Antarctica:

incredible scenery, mountains
fossils, geology
Weddell seals
cooking/meals at Scott Base
helicopter travel
the roar of the primus
"independence" in the deep field
the historic huts
penguins
Mac computers at Scott Base
sunshine on windless days
Scott Base bar on a quiet night
getting in sleeping bags in polar tent
getting mail at Scott Base
24-hour daylight in the field
helos dropping-in in the field

Things I don't like about Antarctica:

getting up to urinate each morning outside the tent
frozen fingers, toes, ears
winds in excess of 20 knots
VXE-6 delays
skidoo thumb syndrome
double climbing boots
heavy packs laden with rocks
static shocks at Scott Base
defecating outside in cold winds
missing home (family)
waiting around at non-Aztec outcrops
having to right upturned sledges
steep ice slopes
poor visibility conditions
avalanches burying me
falling into crevasses
being tent-bound because of snow

Appendix 2
Some Great Recipe Ideas from Antarctica

Peppermint Chicken à la Staite

Brush your teeth using a peppermint-flavored toothpaste. Spit out the rinse water into a clean fry pan (it must be spic and span, hygiene is of utmost importance), then leave it outside in –4°F to freeze (if you are not in the Antarctic, put it in your freezer). Take a frozen chicken from the food box and thaw it slowly by heating it in a camp oven with some hot water, then cut it into pieces. Fetch your fry pan from outside with the frozen toothpaste water and cook the chicken pieces in it, adding some oil, salt, and pepper to taste. Gives a delicious peppermint flavor. Margaret and Fraka both commented on how nice it was, but Brian and I would not reveal our secret ingredient to them.

Deception Irish Cream

Named after the Deception Glacier where we invented the recipe, and the fact that we were deceiving the drinker into thinking it was a real brand of Irish Cream liqueur.

Take a cheap and nasty brand of whisky. Ours was a particularly good value drop then available from the Scott Base bar that sold for about $NZ10 a liter!

Mix up about 2 cups full of full cream milk powder with about half the correct amount of water, thus making "mock cream." Add about a half a cup of sugar dissolved in a little hot water and milk powder, then mix it all up with about 500 ml of whisky. Beautiful, can't be distinguished from any "genuine" Irish Cream (at least to us out on the glacier) but costs next to nothing to make.

Deception "Kar-lua"

Basically similar to Deception Irish Cream except that we mixed about five teaspoons of coffee with our powdered milk and sugar, dissolved it all in a slurry of hot water and added it to a good dose of cheap rum. Makes an excellent drink.

References and Source Material

Select references: historical and literary

Aagaard, B. *Fangst og Forskning i Sydishavet* (Oslo, 1930).
Amundsen, R. *The South Pole* (John Murray, London, 1912).
Barber, N. *The White Desert* (Hodder & Stoughton, London, 1958).
Bowden, T. *Antarctica and Back in Sixty Days* (Australian Broadcasting Company, Sydney, 1991).
—— *The Silence Calling: A History of Australians in Antarctica* (Allen and Unwin, Sydney, 1998).
Bull, H. *The Cruise of the Antarctic* (Blurtisham Books/Paradigm Press, Suffolk, 1984; first published 1896).
Byrd, R. *Discovery* (Putnam, New York, 1935).
—— *Alone* (Putnam, New York, 1939).
Cherry-Garrard, A. *The Worst Journey in the World* (Chatto and Windus, London, 1922).
Cook, J. *The Voyages of the Resolution and Adventure 1772-75* (Hakluyt Society, London, 1961) (ed. by J.C. Beaglehole).
Evans, E.R.G.R. *South with Scott* (Collins, London & Glasgow, 1931).
Gemmell, N. *Shiver* (Vantage Books, Australia, 1997).
Jacka, F. & Jacka, E. (eds.) *Mawson's Antarctic Diaries* (Allen and Unwin, Australia, 1988).
Lee, C. *Snow, Ice and Penguins* (Dodd Mead, New York, 1950).
Lovecraft, H.P. *At the Mountains of Madness, and Other Novels of Terror* (Panther Books/ Granada, London, 1968).
Mawson, D. *Home of the Blizzard* (Heinemann, London, 1930; single volume condensed version, reissued 1996).
Mear, R. & Swann, R. *In the Footsteps of Scott* (Jonathan Cape, London, 1987).
Nansen, F. *Farthest North* (Macmillan, London, 1897).

Nordenskjold, O. *Antarctica, or Two Years Amongst the Ice of the South Pole* (Macmillan, New York, 1905).
Quartermain, L.B. & Markham, G.W. "New Zealanders in the Antarctic," *New Zealand Journal of Geology and Geophysics* 5 (1962): 673-80.
Scott, R.F. *Scott's Last Expedition* (Smith, Elder, London, 1913).
Seaver, G. *Edward Wilson of the Antarctic: Naturalist and Friend* (John Murray, London, 1933).
Shackleton, E. *The Heart of the Antarctic* (Heinemann, London, 1909).
—— *South* (Macmillan, New York, 1926).
Suter, K. *Antarctica: Private Property or Public Heritage?* (Pluto, Australia, 1991).
Wheeler, S. *Terra Incognita: Travels in Antarctica* (Vintage, London, 1997).
Zinsmeister, W.J. "Early geological exploration of Seymour Island, Antarctica," *Geological Society of America, Memoir* 169 (1988): 1-19.

Source materials: some scientific papers

Bradshaw, M.A. "Paleoenvironmental interpretations and systematics of Devonian trace fossils from the Taylor Group (lower Beacon Supergroup), Antarctica," *New Zealand Journal of Geology and Geophysics* 24 (1981): 615-652.
Bradshaw, M.A., Harmsen, F.J. and Kirkbride, M.P. "Preliminary result of the 1988-89 expedition to the Darwin Glacier area," *New Zealand Antarctic Record* 10 (1990): 28-48.
Colbert, E.H. *Men and Dinosaurs* (Penguin, Middlesex, UK, 1968).
Gasparini, Z., Olivero, E., Scasso, R. & Rinaldi, C. "Un ankylosaurio (Reptilia, Ornithischia) Campaniano en el continente Antarcttico" *Anais do X Congresso Brasileiro de Paleontologia* 13 (1987): 131-141.
Hammer, W.R. & Hickerson, W.J. "A crested theropod dinosaur from Antarctica," *Science* 264 (1994): 828-830.
Hampe, O. & Long, J.A. "The histology of Middle Devonian chondrichthyan teeth from southern Victoria Land, Antarctica," *Records of the Western Australian Museum Supplement* no. 57 (1999): 23-36.
Hill, R.S., "*Nothofagus* Fossils in the Sirius Group, Transantarctic Mountains: Leaves and Pollen and Their Climatic Implications," Antarctic Research Series—*American Geophysical Union* 60 (1993): 67-74.
Hooker, J.J., Milner, A.C. & Sequiera, S.E.K. "An ornithopod dinosaur from the Late Cretaceous of West Antarctica," *Antarctica Science* 3 (1991): 331-332.
Long, J.A. "New fossil fish discoveries from South Victoria Land help correlate Devonian sequences in Australia," *Anare News* 70-71 (1993): 22-24.
—— *The Rise of Fishes—Their 500 Million Year History* (University of New South Wales Press, Sydney & Johns Hopkins University Press, Baltimore (1995).
——"A new groenlandaspidid arthrodire (Pisces; Placodermi) from the Middle Devonian

Aztec Siltstone, South Victoria Land, Antarctica," *Records of the Western Australian Museum* 17 (1995): 35-41.

Long, J.A. & Young, G.C. "Sharks from the Middle-Late Devonian Aztec Siltstone, southern Victoria Land, Antarctica," *Records of the Western Australian Museum* 17 (1995): 287-308.

McLoughlin, S. & Long, J.A. "New Records of Devonian Plants from southern Victoria Land, Antarctica," *Geological Magazine* 131 (1994): 81-90.

Ritchie, A. "Ancient animals of Antarctica—part 2," *Hemisphere* 15, Part 12 (1971a): 12-17.

—— "Fossil fish discoveries in Antarctica," *Australian Natural History* 17 (1971b): 65-71.

—— "*Groenlandaspis* in Antarctica, Australia and Europe," *Nature* 254 (1975): 569-573.

Thomson, M.R.A. "Antarctic invertebrate fossils: the Mesozoic-Cainozoic record," in *The Geology of Antarctica*, edited by R.J. Tingey (Oxford University Press, Oxford, 1989): 487-498.

Turner, S. & Young, G.C. "Thelodont scales from the Middle-Late Devonian Aztec Siltstone, southern Victoria Land, Antarctica," *Antarctic Science* 4 (1992): 89-102.

White, E.I. "Devonian fishes of the Mawson-Mulock area, Victoria Land. Antarctica, Trans-Antarctic Expedition 1955-1958, Scientific Reports," *Geology* 16 (1968): 1-26.

Wiman, C. "Über die alttertiären Vertebraten der Seymourinsel," *Wissenschaftliche Ergebnisse der Schwedischen Säudpolar-Expedition 1901-1903* 3 (1905): 1-37.

Woodburne, M.O. & Zinsmeister, W.J. "The first land mammal from Antarctica and its biogeographic implications," *Journal of Paleontology* 58 (1984): 913-948.

Woodward, A.S. "Fish remains from the Upper Old Red Sandstone of Granite Harbour, Antarctica. British Antarctic (Terra Nova) Expedition 1910, Natural History Report," *Geology* 1 (1921):1-62.

Woolfe K.J., Long, J.A., Bradshaw, M.A., Harmsen, F. & Kirkbride, M. "Fish-bearing Aztec Siltstone (Devonian) in the Cook Mountains, Antarctica," *New Zealand Journal of Geology and Geophysics* 33 (1990): 511-514.

Young, G.C. "Devonian sharks from south-eastern Australia and Antarctica," *Palaeontology* 25 (1982): 817-843.

—— "Antiarchs (placoderm fishes) from the Devonian Aztec Siltstone, southern Victoria Land, Antarctica," *Palaeontographica A* 202 (1988): 1-125.

—— "New occurrences of culmacanthid acanthodians (Pisces, Devonian) from Antarctica and south-eastern Australia," *Proceedings of the Linnean Society of New South Wales* 111 (1989): 11-24.

—— "Fossil fishes from Antarctica," in *The Geology of Antarctica*, edited by R.J. Tingey (Oxford University Press, Oxford, 1991)): 538-567.

Young, G.C., Long J.A. & Ritchie, A. "Crossopterygian fishes from the Devonian of Antarctica: systematics, relationships and biogeographic significance," *Records of the Australian Museum, Supplement* 14 (1992): 1-77.

Young, V.T. "Early Devonian fish material from the Horlick Formation, Ohio Range, Antarctica," *Alcheringa* 10 (1986): 35-44.

Acknowledgments

My two scientific expeditions to Antarctica were funded though grants from the National Geographic Society of America (1988-89, Grant #3895-88) and the Australian Antarctic Division (1991-92, ASAC Grant #136). To both of these organizations I give my deepest thanks, not only for the scientific results that we achieved, but also for the opportunity to experience this amazing continent. I deeply thank my expedition mates on the 1991-92 trip, Margaret Bradshaw, Fraka Harmsen, and Brian Staite, not only for their constant support and companionship, but also for their good humor, tolerance, and zest for life. Margaret Bradshaw got the ball rolling for me to participate in both expeditions, and for this, Margaret, I am eternally grateful to you, as well as for the hospitality you and John showed me when I was in New Zealand. There were many people from the NZARP who made my two stays at Scott Base and Vanda Station so memorable; I thank them all from the bottom of my heart for their friendship and guidance, particularly Phil Robbins, Dave Carrera, John Alexander, and Dave Geddes. Thanks also to Jenny and Alex Bevan, who loaned me their now dog-eared but much revered copy of *At the Mountains of Madness.* For reading and commenting on the draft manuscript, thanks to Kerry Bawden, Kim Hudson, Ian Bowring, and the editorial team at Allen and Unwin.

Last but not least, thanks to my three children, Sarah, Peter, and Madeleine, who gave me the inspiration and motivation to put all this down on paper. To use a line from my favorite film, *Conan the Barbarian*, "You are the wellspring from which I flow."

Index

A factor, 51, 194
Aagaard, 12
Aberdeen University, 152
Acanthodian fishes, 17-18, 120, 134, 143, 145, 146, 196, 199
Access to deep-field areas, 7-8, 147
Acy Deucy bar, 52-53
Adelie Land, 73-74, 95
Adelie penguins, 75-77, 78
Adolphspoort formation (S. Africa), 18
Africa, continental drift, 3
Ahlberg, Per, 176-177
Airdevron 6 ice falls, 86
Airplane crash, 32
Alexander, John, 37, 39
Alexander Terrane, 199
Allen, Scott, 88
Allibone, Andrew, 38, 50, 58, 60, 62, 65
Alligator Peak, 143, 166, 167, 168, 169, 170
Alligator Ridge, 166, 168-171, 172
Allosaurus, 17
Alone (book), 215
Amphibian fossils, 176-177
Amundsen, Raold, 6, 11, 14, 31, 78
Anachronistes, 199
ANARE (Australian National Antarctic Research Expeditions), 68-69, 124, 216
Anareodus staitei, 124
Ankylosaur, 16, 86
Antarctaspis mcmurdoensis, 179
Antarctic (ship), 5, 13, 99
Antarctic Circle, 5
Antarctic Conquest (book), 42-43
Antarctic Development Squadron Six (VX-6), 30-31
Antarctic Heritage Trust, 37
Antarctic Peninsula, 13, 16, 47, 86
Antarctic Times (newsletter), 48-49
Antarctic Treaty, 19, 221
Antarctica
- access to deep-field areas, 7-8, 147
- airplane crash, 32
- Australian Territory, 7, 68
- biomass of, 36, 220-221
- cars in, 6
- dinosaurs in, 16-17, 82, 86, 113
- discovery of, 2, 4, 5
- electric shocks (static electricity) in, 35
- exhibition, 216
- fish species, 36, 220-221
- forests, 3-4
- formation of, 2, 3
- fossils found in, 12, 12, 13-14
- fresh water, 19, 220

human habitation of, 219-221
likes and dislikes about, 227-228
mythological background, 4-5
natural resources, 19-21, 219-220
nuclear power in, 35-36
sand dunes in, 61
size of, 18-19
tourism, 20-21, 32, 221-222
women in, 9-10
Antarctica and Back in 60 Days (book), 165
Antarctilamma prisca, 194, 197
Antarctos, 4
Arabia, 3
Archaean rocks, 2
Archaeocyathids, 13, 46
Archaeosigillaria, 103
Arctos (star), 4
Argentina, 16, 103
Aristotle, 4
Arresting technique, 39-40, 42
Art exhibition, 37
Arthrodiran placoderms, 123
Arthropods
giant, 152, 153
walking on dry land, 153
Asgaard Range, 50, 58, 62
Asgaard Rangers, 63
Augustana College, 17
Aurora (ship), 73-74
Australia
Antarctic Division, 68
Antarctic Territory, 7, 68
continental drift rate, 3
fossil fish fauna, 8, 118, 120, 147, 179, 198, 199-200
Gondwana and, 3, 8, 16, 47, 133, 147, 160, 170, 216
plant fossils, 16
Tumblagooda Sandstone, 152-153
women in Antarctica, 10
Australian National University, 8, 59
Australasian Antarctic Expedition (1911-14), 7, 45, 46, 82, 129
Australian Museum, 8, 175
Austrophyllolepsis, 118, 122
Avalanches, ix, 186, 187, 223
Ayres, Harry, 110
Aztec Siltstone, 66
on Alligator Ridge, 166, 168
in Boomerang Range, 157, 158
calcareous layers, 196
characteristics of, 83, 104, 116, 132
in Cook Mountains, 65, 92
dating of, 170
Devonian fish fossils, 14, 18, 50, 104, 160, 170, 199-200
discovery in Antarctica, 92
in Dry Valleys, 64
at Fault Bluff, 117-118, 120, 125
on Gorgon's Head, 102, 104, 106
measurements, 134, 146
on Mt. Crean, 196, 197-198
on Mt. Gudmundson, 132, 133, 134
on Mt. Ritchie, 142, 146
origin of name, 9
at Portal Mountain, 178, 190
on Seay Peak, 138
shark fossils in, 198
Aztecodus harmsenae, 197-198

Bach, Richard, 127, 129
Bages, 164
Balaena (ship), 12
BANZARE (British-Australian-New Zealand Antarctic Research Expedition), 1, 7
Barbecues, 40-41, 44
Barber, Noel, 23, 80
Barnes Glacier, 81
Barrett, Peter, 108
"Barrier Silence," 110
Barrydale-Ladismith region (S. Africa), 170
Barwick, Richard E., 59
Barwick Valley, 59
Bathing, 157
Baynes, Alex, 164
Beach parties, 34
Beacon Heights, 107-108

Beacon Heights Orthoquartzite, 125, 133, 153, 158
Beacon Supergroup, 50, 59, 85, 87, 96, 105-106, 108, 113
Beaconites, 107-108, 117, 125, 135, 153, 170
B. antarcticus, 107, 108
B. barretti, 106, 107, 108
Beardmore Glacier, 7, 13, 14, 15-16, 91
Beelarongia, 161
Berlin Natural History Museum, 189
Biomass of Antarctica, 36, 220-221
Black Island, 34
Blizzards, 1, 43, 88
in Adelie Land, 95
deaths during, 15
during deep-field expeditions, 92-94, 156, 162
health concerns, 2, 15, 195
meteorological conditions, 95, 97, 98, 158, 192, 195
on polar plateau, 2
sledging in, 92-94
tent-bound activities during, 156-163, 194, 195
tent erection in, 24, 94
Blubber stove, 27
Blue ice, 92, 138, 139, 157-158, 160, 174
"Boomerang" flights, 32
Boomerang Range, 49, 50, 66, 146, 151, 155, 156, 157, 162, 166, 170, 182, 185, 186
Boomeraspis, 170
Borchgrevink, Carsten, 5-6
Bothriolepis, 86-87, 119-120, 123, 124, 125, 134, 170, 175, 189, 197
B. mawsoni, 145
Bowden, Tim, 165
Bowers, 7, 43, 188
Bradshaw, John, 10
Bradshaw, Margaret, 166, 177, 204, 209, 210
cooking skills, 105, 111, 139, 151, 164, 165, 174, 191
cross-country skiing, 66
on Darwin Glacier, 44, 48-49, 50, 65, 89, 165, 202
to Escalade Peak, 150
expeditions, 9, 10-11, 23, 38, 68, 165
expertise, 10, 88, 105-106, 108, 124, 135
at Fault Bluff, 117, 124, 126
fossil collections/discoveries, 9, 105-106, 107, 109, 135, 152, 170
at Gorgon's Head, 93, 104, 105-106, 107, 109
on McCleary Glacier, 111
on Mt. Gudmundson, 133, 134
on Mt. Ritchie, 145
on Mulock Glacier, 139
personal characteristics, 9, 105, 119, 179, 190
at Portal Mountain, 182, 189
species named for, 189
British Antarctic Expedition, 6, 75, 77
British Antarctic Survey, Marine Life Sciences Division, 221
British Natural History Museum (London), 15, 176-177, 179, 198, 199, 215
Brown, Harold, 131
Buckley Island, 188
Bull, Henryk, 5-6
Bull Pass, 60, 61
Butson, 42-43
Byrd Polar Research Center, 17
Byrd, Richard (Admiral), xiv, 172, 173, 187, 215
Byrd, Robert, 12

Calcareous nodules, 196
California State University at Fresno, 23
Cambrian fossils, 13, 16, 46-47
Campbell, Ken S., 59
Campbell, Victor, 27, 76
Canada, 199
Canowindra (NSW), 160
Canowindra, 160-161, 176
Canterbury Museum, 9, 10
Canterbury University, 10

Cape Adare, 5-6
Cape Byrd, 43
Cape Crozier, 188
Cape Denison, 95, 203
Cape Evans, 75, 78-80
Cape Royds, 75, 76, 78
Carboniferous deposits, 124, 199
Carlyon, Roy, 110
Carrera, Dave, 33
Cars in Antarctica, 6, 77
Cedarberg Mountains (S. Africa), 18
Chalet mobile house, 40, 52, 66
Chaplin, George, 217
Cherry-Garrard, Apsley, xii, 43, 177, 188
Christchurch (NZ), 210
  US Naval Air Base, 30
  Windsor Hotel, 29-30
Christmas
  celebrations, 52-55, 159, 161, 162, 164-165
  feasts, 54-55, 155, 164, 165
  gift exchanges, 165
  poem, 163
Chondrichthyans, 200
Churchill Mountains, 90
Circumpolar current, 4
Clichy-Peterson, Harry, 42-43
Climate variations, 19-20
Clothing, protective, 2, 24, 30, 37, 43, 68, 93, 98, 216
Cocktails, 148, 151, 159, 164-165, 174, 179, 190
Coelacanths, 160
Colbert, Edwin H., 16
Commercial ships, 222
Commonwealth Bay, 7
Commonwealth Transantarctic Expedition, 33
Confucius, 22
conodont microfossils, 153
Continental drift, 2-3, 8, 11, 14; *see also* Gondwana
Cook, James, 5
Cook Mountains, xii, 9, 65, 85, 91, 94-95, 96, 110, 125, 134, 136
Corbett, Dennis, 62
Correll, Perry, 73-74
Cosmine, 175-176
Crampons, 41, 105, 146, 186, 196, 197
Crean, 193, 194
Cretaceous Period, 3
Crevasses, ix, 22, 86, 178
  anchoring techniques, 41-42
  detection of, 41, 48, 84, 112, 114, 138, 187
  falls into, 100, 110, 181, 184-185, 187, 190, 214-215, 223
  fields, 88, 89, 92, 111, 112, 114, 151, 169, 172, 190
  formation of, 85, 137-138
  ice bridges, 41, 89
  navigation of, 112, 114, 121, 121, 138, 155, 186, 191, 197
  prussocking out of, 38, 41-42
  rescue/recovery from, 24, 42-43, 101, 110, 181
  and sledging, 138
  small, 105, 121, 138, 197
  view from, 81
Crinoids, 36
Cross-country skiing, 66, 208
Crossopterygian fishes, 160, 175
Cruise of the Antarctic (book), 5-6
Cryogenics, 20
Cryolophosaurus ellioti, 17, 86
Culmacanthus, 146
  C. antarctica, 147

Dais structure, 61-62, 64
Dalziel, Ian, 47
Darlington, Harry, 10
Darlington, Jennie, 10
Darwin, Charles, 11, 90
Darwin Glacier, 38, 48-49, 85, 89-90, 91, 92, 110, 111, 123, 136, 181
Darwin Mountains, 65, 85
Dating of sedimentary sequences/rocks, 2, 18, 122, 152, 170
David, Edgeworth, 6, 13
Deaths of explorers, 7, 15, 183

Debenham, Frank, 14
Deception Glacier, 138, 146, 147-148, 149, 150, 161
Deep-field expeditions, 46
A factor, 51, 194
access to areas, 7-8, 147
adjustment to return from, 205, 206-212, 213
bathing, 157
blizzard conditions, 92-94, 156, 162
cocktails, 148, 151, 159, 164-165, 174, 179, 190
days lost to bad weather, 182
food, 94-95, 105, 108, 109, 114, 144, 145, 146, 148, 150, 151, 155, 161, 164, 190, 191, 202
group dynamics, 126, 139, 165-166, 169, 190
health issues, 14, 22, 24-25, 26, 151, 159, 184, 194, 195, 203
laundry operations, 46, 157
lavatory facilities, 139-141, 162-163, 194, 209
logistic support, 67-68, 72, 83, 108, 116, 135, 136, 147, 150, 156
New Year's celebrations, 179-180
positioning, 8
pullout/pick-up, 191, 195, 196-197, 201-202, 203-205
radio communications, 24, 27, 89-90, 94, 122-123, 162, 164, 180
recon flights, 83, 84-87, 112, 196-197, 200
recreation, 109, 111-112, 119, 125, 156-157, 158, 159, 161, 162, 163, 164, 179-180, 190, 191, 195
retrieval of gear, 205-206
route markers, 84, 85, 189, 157-158
sledging flags, 86-87
survival training, 37-38, 39-41, 69, 72-73
tent-bound activities during, 156-163, 194, 195
women, 9-10
Dental work. *See also* Teeth
preparation for Antarctica, 22-23
self-performed, 24
Derycke Park meteorite, 65
Devonian period, 17
amphibians, 176
bivalves, 10
Early Devonian, 106, 152, 199
fishes, 8-9, 13-14, 14, 18, 50, 104, 160, 170, 199-200
land, 153
sedimentary rock, 59, 120
sharks, 189
trace fossils, 50
Dinosaurs, 3, 195
in Antarctica, 16-17, 82, 86, 113
Diplichnites, 153
Diprotodon, 163
Discovery expedition (1901-1904), 6, 11, 36-37, 58, 90, 99, 107-108
Discovery of Antarctica, 2, 4, 5
Dogs. *See also* Huskies
crevasse accidents, 110
as food, 100, 155
Dolfin fossils, 16
Doliodus, 199
Dolorite, 113, 114, 127, 148, 174, 178, 184
Don Juan Pond, 64
Don the Magician, 53
Doyle, Martin, 58, 60, 62
Drinking water, 2
Dry Valleys, 49-50, 56, 57-62, 178
Durmont D'Urville, 65-66

Early Devonian, 106, 152, 199
Early Jurassic sandstones, 17
Early Oligocene epoch, 3
Echos, 81
Electric shocks in Antarctica, 35, 208
Elgin (Scotland), 176-177
Elginerpeton, 176-177
Elliot, David, 17
Emergency caches, 80, 100, 101, 174

Emperor penguins, 43, 76, 79, 188, 216
Enderby Land, 2
Endorsement of brand names, 77
Endurance (ship), 137
Engine maintenance, 24
Environmental protection measures, 2, 19-20, 35-36, 51, 52, 221-222
Erebus (ship), 29
Erebus Ice Tongue, 81
Escalade Peak, 149, 150-154, 155
Euphasia superba, 220
Euramerica, 177
Europe
 continental drift, 3
 Old Red Sandstone fossils, 14
 phlyctaenioid arthrodires, 170
 placoderm fossils, 118
Eurypterid tracks, 152, 153
Evans, Lt., 15, 193-194
Extinction events, 193

Facies, 132
Falloon, Garth, 180
Famennian stage, 170
Farthest North, 209
Fault Bluff, 113, 114, 116-126
Festive Plateau, 86, 110, 111, 113
Field cookery
 corned beef curry, 191
 Deception Irish Cream, 148, 229-230
 dehydrated meals, 62, 105, 109, 145, 150, 159, 194, 202
 desserts, 146, 151, 164, 174, 190
 dog soup, 155
 fresh foods, 146
 hoosh, 100
 Mock Kar-lua, 151, 230
 peppermint chicken, 177
 recipes, 229-230
 roast lamb dinner, 94-95
 sweet-and-sour saveloys, 62
 Thai satay beef with khao padt, 159
Finger Ranges, 135, 137
Fischer Catering, 44
Fish of Antarctica, 36, 220-221; *see also* Fish fossils
Fish fossils. *See also specific genus*
 Australian fauna, 8, 118, 120, 147, 179, 198, 199-200
 castings, 56, 135
 Devonian period, 8, 10, 13-14, 17, 104, 117, 120
 dispersal routes, 120
 first discovery in Antarctica, 13-14
 living, 160
 lobe-finned, 17, 119-120, 160-161, 175-177, 197
 packing of specimens, 124, 135
 ray-finned, 18
 sites, 32, 64, 65, 91, 92, 169, 178
Fish Hotel, 122-125
Fitzgerald, Paul, 122, 180, 205
Flannery, Tim, 36
Floodplain environments, ancient, 168
Food, 2, 142. *See also* Field cookery
 barbecues, 40-41, 44
 box lunches, 31-32, 69
 breakfast, 181-182
 cocktails, 148, 151, 159, 164-165, 174, 179, 190, 229-230
 dehydrated, 62, 145, 150, 159, 194
 dogs as, 100, 155
 dreams of, 45, 46, 100
 early expeditions, 77, 80, 99, 155, 165
 emergency caches, 80, 100, 101, 174
 fresh resupplies, 145, 146, 165, 174
 holiday feasts, 54-55, 155, 164, 174, 179
 hoosh, 100
 meat preservation, 169, 174
 NZARP boxes, 98-99, 106, 174
 pemmican, 99, 100, 101
 penguins as, 27, 76, 99, 220
 product endorsements, 77
 psychological role, 98, 101, 144, 185
 recovery from crevasses, 110
 at Scott Base, 35, 82
 seals as, 27, 37, 52, 76, 99, 220

shortages/rationing, 161, 163
skuas as, 129
sledging rations, 99, 101
Food chain, 36, 220
Forests, Antarctic, 4
Formation of Antarctica, 2, 3
Forsyth, Jane, 38
Fossil-bearing rocks, 9, 11. *See also* Aztec Siltstone; *other specific types*
dating of, 18, 152, 170
fish-bearing horizons, 104
weathering action on, 176
Fossils. *See also* Fish fossils; *specific genus*
erosion of, 143-144
extraction from rock, 134
first found in Antarctica, 12, 13-14
locations of, 8-9, 11, 13, 17, 32, 46, 64, 142-144
uses of, 18-19, 120
Four-corners philosophy, 218
Foyn, Svend, 5
Fram (ship), 78
Frasnian Stage, 170, 177
Fresh water, 19, 220
Frostbite, 14, 24-25, 26, 203
Frostnip, 25, 195
Fuchs, Vivian (Sir), 28
Fuelwood, 40-41

Geddes, David, 140-141, 209
Gemmel, Nikki, 183
Geological Society of Ireland, 215
Gillespie, Grant, 49
Givetian stage, 170
Glacial deposits, 13, 14, 64, 133, 143, 158
Glacial melt rivers, 61
Glacier crossing techniques, 41
Glossopteris, 16
Glycopeptides, 36
Glyptonotus, 36
Goddard, E.J., 13
Gondwana, 11
Australia and, 3, 8, 16, 47, 133, 147, 160, 170, 216
biogeographic links, 8, 16, 18, 47, 103, 118, 160-161, 216
biostratigraphic links, 133, 170
conferences, 68
endemic fish fossils, 176, 179, 199
formation, 2-3
India and, 3, 16, 133
North America and, 3, 47, 177
South Africa and, 16, 133, 216
Gordon, W.T., 13
Gorgon's Head, 91, 92, 96, 97, 102-109, 116, 150, 152
Granite Harbour, 14
Granulitic rocks, 2
Greenland
Devonian rocks, 120
placoderm fossils, 118
Greenpeace Base, 78, 80-81
Groenlandaspis, 119, 120, 124, 170, 179, 196, 197
Group dynamics, 126, 139, 165-166, 169, 190
Gunn, Bernie, 179, 198
Gyracanthides, 118, 120

Hagglunds snowmobile, 39
Hammer, William, 17
Hampe, Oliver, 189
Haplostigma lineare, 103
Harmsen, Fraka, 93, 150, 174, 177, 191, 204, 209-210
Christmas celebrations, 164, 166
cooking skills, 202
expertise, 23, 85, 120-121, 131
at Fault Bluff, 117, 120-121, 124
at Gorgon's Head, 104, 107, 109
on Lashly Glacier, 201
on Mt. Gudmundson, 131, 133
on Mt. Ritchie, 145
personal characteristics, 23, 119
at Portal Mountain, 182, 191
Harrington, Larry, 107
Harrowfield, David, 37
Hatherton, Trove, 106
Hatherton Sandstone, 106, 107, 125, 151-152, 153

Hawke, Bob, 162
Hawkes, William "Trigger," 31
Health issues. *See also* Psychological challenges
blizzards, 2, 15, 195
on deep-field expeditions, 14, 22, 24-25, 26, 151, 159, 184, 194, 195, 203
dehydrated foods, 194
diary entry, 151
frostbite and frostnip, 14, 24-25, 26, 195, 203
hypothermia, 25-26, 99
medical checkups, 22-23
medical emergencies, 26
nutritional/metabolic, 6, 45, 46, 61, 100-101, 184, 194, 213—214
personal hygiene, 24, 151, 157, 212
snow blindness, 159
teeth, problems with, 23, 71
Heimdall Glacier, 63
Heimdallia, 62, 153
Helicopters, 31, 63, 65
area accessible with, 7
crashes, 65-66
dangers of riding in, 59
familiarization course, 26
food/supply drops, 156e
landing on ice, 89
loading and unloading, 88, 89, 137, 138
NZ01, 156
skidoo transport, 135
Henare, John, 49
Hercules (C-130), 31, 182
areas accessible by, 7-8
Cargo Cult, 201
landing hazards, 30, 48, 200-201, 204
recon, 196-197
recovery of XD-03, 65-66
runway areas, 201—202, 203-204
XD-01, 203
XRD-04, 202
Hickerson, William, 17
Hillary, Edmund, 28
Hills, Edwin Sherbon, 118
Historic huts, xiii, 15, 36-37, 75, 77, 78-80
Holocephalans, 200
Homesickness, 24, 54, 97, 126, 156-157, 165, 188
"Honey pot", 58
Hoosh, 100
Hope Bay, 13
Hornibrook Symposium, 9
Horseshoe crab, 107
Human habitation of Antarctica, 219-221
Huskies, 51-52, 59, 77-78
Hut Point, 36
Hydroponic agriculture, 220
Hypervitaminosis A, 100-101
Hypothermia, 25-26, 99
Hypsilophodontids, 16

Ice bridges, 41, 89
Ice caps, 4
Ice casts, 135
Ice caves, 27, 40, 41, 76, 81-82
Ice falls, 85, 86, 111, 112, 113
Ice floes, 81
Ice ridges, 133
Ice rivers, 112, 113
Ice sheet, Antarctic, 19, 34, 38, 58
Ice shelf, 57
Ice tongues, 32, 81
Ice walls, 59, 81
Ice waves, 121
Icebergs, 77, 78
Igloos, 40, 41, 81
Immersion foot, 24
India, continental drift, 3, 16, 133
Indonesia, living fossil fishes, 160
International Geophysical Year (IGY), 10, 33, 35, 58, 198
Iran, 216
Ireland, placoderm fish fossils, 120
Isolation, 24, 97
isopod, giant marine, 156

James Cook University, 92
James Ross Island. *See* Ross Island
Jason (ship), 12
Johnston, David, 215
Joly, 4
Jonathan Livingston Seagull, 127, 129
Junction Spur Sandstone, 96, 107, 150
Jurassic plant fossils, 13

Kalbarri (W. Australia), 108, 152
Kanak Peak, 126, 128
Karate, 217-218
Karratha, 163-164
Keating, Paul, 162
Kenyite, 81
Killer whales, 57
Kiwis, 44, 53, 71, 162
Klipbokkop formation (S. Africa), 18
Koharolepis, 160-161, 176
Kohn, Barry, 147
Krill, 220-221

Labyrinth, 84
Land bridges, 3
Lake Tekapo (NZ), 23-24
Lake Vanda, 55
  characteristics, 58-59, 64, 84
  swim club, 63-64, 72
Lake Vashka, 59
Lake Victoria, 50, 58, 59, 60
Lake Vida, 59
Larsen, Carl, 12
Lashly, 193, 194
Lashly Glacier, 191, 192, 201, 205
Lashly Ranges, 191, 192, 193, 194, 197
Late Devonian, 103, 118, 170, 198, 200
Late Eocene epoch, 3
Late Ordovician age, 153
Latimeria, 160
Laundry operations, 46, 157
Lava tongues, 33
Lavatory facilities, deep-field, 70, 139-141, 162-163, 194, 209
Leonodus, 199
Limestone fossils, 196
Logistic support for expeditions, 67-68, 72, 83, 108, 116, 135, 136, 147, 150, 156
Lord Howe Rise, 3
Lord of the Rings, 126
Lovecraft, H.P., xii, 1, 4, 57, 102, 109, 125, 142, 149, 168, 193, 207, 213
Lungfishes, 17, 196
Lycophytes (lycopods), 103, 120

Mackay Glacier, 14
Malanzania, 103
Marsupial fossil, 163
Massive Attack, 223
Mawson, Douglas
  Australasian Antarctic Expedition (1911-14), 7, 45, 46, 82, 129
  BANZARE (1929-31), 1, 7
  base area, xiii, 95
  on blizzards, 88, 95
  British Antarctic Expedition (1907-09), 6
  crevasse accident, 181
  food/starvation, 45, 46, 99, 100-101, 155, 165, 203
  emperor penguin, 216
  health, 203
  Polar Plungers Club, 73-74
  sledging rations, 99
  species named for, 145
McCleary Glacier, 84, 85, 86, 89, 108, 110-115
McCormick, Robert, 29
McKelvey Valley, 60
McLoughlin, Steve, 102
McMurdo district fossils, 16
McMurdo Base (Mactown), 23, 32, 35-37, 44, 46, 52-54, 56, 81, 86, 147, 178
McMurdodus featherenesis, 198-199
McNamara, Ken J., 108, 163
Medical emergencies, 26
Medical tests for Antarctica, 22-23
Melbourne University, 118
Men and Dinosaurs (book), 16

Mertz, Xavier, 7, 100, 101
Metaxygnathus, 176
Meteorites, 65, 216
Meteorology of Antarctica
conditions associated with blizzards, 95, 97, 98, 158, 192, 195
recordings, 63
temperature extremes, 116, 118-119, 120, 122, 134, 135, 142, 150, 151, 194, 202
wind-chill factors, 82, 87, 96, 156, 194
wind speeds, 57-58, 79, 82, 92, 95, 97, 145, 147, 160, 164, 166, 194, 195
Middle Devonian, 177, 198, 199, 200
Middle East, 3, 120
Mikkelsen, Caroline, 10
Mollusk fossils, 46, 47
Monash University, 215
Moody Peak, 148, 149
Moreton Bay bugs, 135
Morgan, Gary, 156
Mountain climbing, 24, 38, 133-134, 196
Mt. Adam, 122
Mt. Ayres, 110
Mt. Crean, 85, 176, 192, 193, 194-196
Mt. Erebus, 6, 32, 33, 56, 75, 78, 81
Mt. Feather, 198
Mt. Fleming, 64, 84, 85, 177-178, 195
Mt. Gudmundson, 107, 126, 131-136
Mt Howitt (Australia), 118, 120, 122, 146-147, 161, 170
Mt. Hughes, 91, 110
Mt. Kirkpatrick, 17, 86, 91
Mt. Kohn, 146, 147
Mt. Longhurst, 91, 96, 110
Mt. Metschel, 85, 160, 173-177, 190, 195
Mt. Nickell, 59
Mt. Odin, 62, 63
Mt. Ritchie, 123, 138, 142-144, 147
Mt. Suess, 14
Mudstone, 175, 189, 197
Mulock Glacier, 85, 110, 112-113, 125, 126, 128, 135, 136, 137-141, 202
Mummified seals, 61
My Antarctic Honeymoon (book), 10
Mythological background of Antarctica, 4-5

Nansen, Fridtjof, 67, 209
National Geographic Society of America, 11, 68
National Science Foundation (US), 52, 64-65
Natural resources of Antarctica, 19-21, 219-220
Nematodes, 20
Neoselachians, 193, 199
Neptune, Joseph Entrains, 31
New Mountain Sandstone, 62
New South Wales, 176
New Year's celebrations, 63-64, 179-180
New Zealand. *See also* NZARP
Air Force, 26, 58, 200
Antarctic Division, 30
gear, 68
Science Department, 58
Night travel, 171, 172-173
Nimrod expedition, 13
Ninnis, Lt., 7, 100
Nordenskjöld, Otto, 12-13, 99
North America
continental drift, 3, 47
phlyctaenioid arthrodires, 170
placoderm fish fossils, 120
Notorhizodon, 160
Notothenia, 36
Nuclear power in Antarctica, 35-36
NZARP (New Zealand Antarctic Research Program)
expeditioners, 48, 59
field leaders, 71
food supplies, 98-99, 106, 174
funding for expedition, 11
group K5076, 122-123, 180, 205
medical assessments of expeditioners, 22—23
1990-91 season, 67-68
Scott Base administration, 55
visits to deep-field expeditions, 140-141

Oates, Lawrence "Titus," 183
Observation Hill (Cape Evans), 80
Observation Hill (McMurdo), 44
Odin Valley, 62, 63
Okinawa, 217
Old Red Sandstone, 14, 152
Olympus Range, 50, 58
On Origin of Species (book), 11, 90
One Ton Depot, 7, 15
Onyx River, 59, 61
Operation Deep-Freeze, 31
Ordovician rocks, 62
Ore bodies, 132
Ornithopod fossils, 86
Osteolepidids, 120
Osteolepiformes, 161, 176
Otoliths, 118
Oxygen isotope ratios, 20
Ozone layer, hole, 20

Pack ice, 13, 76, 137
Paintin, Ian, 34, 35, 37
Palaeogeography, 47. *See also* Gondwana
Palaeosols, 168
Paleozoic Era, 143
Palmer, Nathaniel, 5
Panderichthyids, 176, 177
Parahelia, 172, 173
Paulett Island, 13, 99
Pemmican, 99, 100, 101
Penguins, 5, 40, 57
  Adelie, 75-77, 78
  emperor, 43, 76, 79, 188, 216
  as food, 27, 76, 99, 220
  fossils, 13
Pensacola Mountains, 47
Permian ice age, 124, 133, 143
Permian Sandstones, 145
Perseus Peak, 152
Personal hygiene, 24, 151, 157, 212
Phlyctaenioid arthrodires, 170, 179
Phyllolepids, 117-118, 120, 122, 145, 179
Phyllolepis, 118
Piss bottles, 140-141, 209
Placoderms, 17, 86-87, 117-118, 120, 122, 134, 170, 179, 196
Plant fossils, 13, 15-16, 102-103, 120, 153
Plasmon bisquits, 99
Plate tectonics concept, 2-3, 4, 18, 200
Plecostomus, 119
Poisonous lakes, 62, 64
Polar plateau
  blizzards, 2
  expeditions crossing, 110, 129
  gateway to, 178
  human habitation of, 220
  katabatic winds, 2, 95, 192, 219
  pollution measurements on, 20
  size of, 7
  view of, 70, 111, 113, 125-126, 134
Polar Plungers Club, 72-73
Polar shelf seas, 36, 220
Polar tents, 27
  emergency, 96
  erection during blizzards, 24, 94
  layout/living conditions, 60, 216
  remains of prior expedition, 178, 195
  securing with ice screws, 174
  temperature inside, 122
Pollution measurements, 20
Ponder, Frank, 33
Ponies, 77-78, 79
Ponting, Herbert, 75, 79, 81
Porolepiformes, 161
Port of Beaumont, The (ship), 10
Portal Mountain, 85, 160, 178, 180, 181-192, 198
Portalodus bradshawae, 189, 190, 198
Potter, Noel, 30, 35
Preparation for expeditions, 67-74
Pressure ridges, 51
Priestley, Raymond, 13
Primus stoves, 24, 27, 40, 94, 122, 157, 163, 174
Private John R. Trowle (ship), 131
Problematica, 46

Propteridophytes, 103
Prussocking, 38, 41-42
Psychological challenges
  adjustment to return from field expeditions, 205, 206-212, 213, 214-215
  food and, 98, 101, 144, 185
  group dynamics, 126, 139, 165-166, 169, 190
  homesickness, 24, 54, 97, 126, 156-157, 165, 188
  humorous response to, 200-201
  isolation, 24, 97
  physical condition and, 217
  privacy issue, 72
  wind, 168
Ptolemy, 4

Quartz, 132
Queensland Museum, 198
Que Sera Sera (aircraft), 31

Rabbit fishes, 200
Radio communications
  Christmas messages, 164
  importance of, 89-90, 162
  New Year's messages, 180
  setup, 89, 94
  training, 24, 27
  with NZARP group K5076, 122-123
Radiometric dating of rocks, 2, 122
Rappelling (abseiling), 42
Razorbacks, 132
Recon flights, 83, 84-87, 112, 196-197, 200
Rescue/recovery
  from crevasses, 24, 42-43, 101, 110, 181
  deep-field pullout/pick-up, 191, 195, 196-197, 201-202, 203-205
  of gear, 205-206
Recreation on deep-field expeditions, 109, 111-112, 119, 125, 156-157, 158, 159, 161, 162, 163, 164, 179-180, 190, 191, 195
Redlichia, 47
Reefs, 20, 46-47
Resting traces, 107, 135
Restoration of historic sites, 37, 77, 78
Retro, 136, 147, 150
R5D aircraft, 31
Rhizodontids, 120
Ritchie, Alex, 8-9, 120, 143, 147, 160, 175-176, 179, 185, 189
Roadend Nunatak, 85
Robbins, Phil, 191
Rock, Nicholas, 214
Ronne, Finne, 10, 42
Ross, James (Captain), 29
Ross Island, xiii, 16, 33, 36-37, 39, 56, 72, 86, 123
Ross Sea, 34, 38, 77
Route markers, 84, 85, 189, 157-158
Rowell, Albert, 46-47
Royal New Zealand Air Force, 137
Royal Society Range, 84
Rudge, Chris, 58, 60, 62, 63-64, 65
Russian explorers, 5, 10, 26

Sand dunes, 61
Sastrugi, 121, 128, 149, 150, 155-156, 157, 166, 173
Saudi Arabia, 199
Saxby, Eric, 66
Scotland
  amphibian fossils, 176-177
  Old Red Sandstone fossils, 14, 152
  placoderm fossils, 118
Scott, Robert Falcon, xiii, 31
  death, 7, 15
  diary entries, 14-15, 39, 75, 100, 160
  Discovery expedition (1901-1904), 6, 11, 36-37, 58, 90, 99, 107-108
  fossil discoveries, 14-16
  geologizing, 132
  memorials, 38, 44, 78, 80
  personal characteristics, 78
  Terra Nova expedition (1910-1913), xii, 6-7, 11, 14, 27, 39, 43, 76, 129, 177, 193-194

Scott Base
  appearance, 32, 33, 56
  bar, 34, 38, 44, 48, 52, 53, 55, 86, 206, 208
  beach parties, 34
  food, 35, 82, 155
  Greenpeace and, 81
  gym, 48, 52
  historic sites, xiii, 15, 36-37, 78-80
  library, 46
  living quarters, 71-72
  post-expedition duties, 208, 209
  radio communications with expeditioners, 122-123, 159, 162, 174, 180
  recreation, 66, 208
  science lab, 82
  wintering over on, 33, 72
Scott Base Times, 82, 208
Scott of the Antarctic (film), 1-2, 70
Scott Polar Research Institute, 15
Scott's huts, xiii, 15, 36-37, 78-80
Scottish Antarctic Expedition, 13
Scree, 63, 102, 104, 105, 142, 143, 158
Scurvy, 6, 194
Sea ice, 51, 78, 82, 99
Sea levels, thawing of Antarctic ice sheet and, 19
Sea lilies, 36
Sea scorpions, giant, 152, 153
Sea urchin fossils, 216
Seals, 5, 34, 40, 57
  conservation of, 221
  as food, 27, 37, 52, 76, 99, 220
  mummified remains, 61
  Weddell, 51, 52, 73
Seay Peak, 125-126, 127-130, 135
Sedimentary rocks, 23. *See also specific types*
  dating of, 2, 18, 122, 152, 170
Sedimentary succession, 153
Seward, A.C., 15-16
Sewell, Rod, 38
Sewing and sewing kits, 68-69, 160
Seymour Island, 12, 13, 216
Shackleton, Ernest, xiii, 6, 13, 15-16, 78, 137
Shackleton's hut, 75, 77
Shark fossils, 17, 18, 123-124, 134, 143, 145-146, 179, 189, 193-200
Shell assemblages, 16
Shinn, Gus, 31
Shiver (book), 183
"Shouting the bar," 48, 208
Silurian Period, 103
Sinfonia Antarctica (musical score), 1-2
Size of Antarctica, 18-19
Skeleton Glacier, 178
Skeleton Névé, 8, 65, 84, 85, 118, 136, 145, 149, 155, 160, 163, 172, 173, 178, 186, 189, 198
Skidoos and skidooing. *See also* Sledging
  air transport of, 88, 135
  fuel, 108, 163
  ideal conditions, 75
  maintenance/protection of, 86, 94
  runway preparation with, 201-202, 203-204
  with sledges, 92, 112, 121, 128, 138-139
  on slopes, 127, 132, 138-139
  speeds, 82, 128, 150
  storage compartments, 128
  wind chill, 82, 87, 127
Skolithos, 107
Skottsberg, C.J., 99
Skua gulls, 129-130
Skua Ridge, 129-130, 131, 150
Sledging
  accidents, 149, 155-156, 166
  in blizzard conditions, 92-94
  on blue ice, 92, 139
  braking, 113-114, 138-139
  in crevassed areas, 138
  with dogs, 6, 51-52
  first major journey, 12-13
  flags, 86-87
  man-hauling, 203
  perfect conditions for, 150, 173

poem, 110
with ponies, 6
rations, 99, 128
relays, 147-148, 191
repairs, 156
route mapping, 84, 85, 189
over sastrugi, 121, 128, 149, 150, 155-156, 166
with skidoos, 92-94, 112, 121, 128, 138-139, 149-150, 155
on slopes, 9, 112, 113-114, 138-139, 147-148, 191
stamina required for, 175
wind speeds and, 92
Sleeping bags, 96
Sleeping outside, 60
Smith-Woodward, Arthur, 14
Snap-frozen animals, 11
Snow blindness, 159
Snow caves, 27
Snow Hill Island, 13
Solar power, 219-220
Sounds of Antarctica, 1-2, 168, 183-184
South Africa
continental drift, 16, 133, 216
fossil fish fauna, 8, 18, 120, 160, 170
South America, 3
South Island (NZ), 22-28
South Magnetic Pole, 6
South Pole, 6, 11, 13, 14, 31, 69, 78
South with Scott (book), 194
Spain, 199
Spiritual/emotional experiences, 29, 172, 173, 223
Spitzbergen, 170
Sponsor's Bluff, 50, 59
Sponsor's Peak, 50
Staite, Brian, 85, 93, 150, 164, 173, 204, 209, 210
at Fault Bluff, 118
field experience/skills, 112, 114, 138, 139-140, 155-156, 163, 181, 182, 187, 196, 201
at Gorgon's Head, 105, 109
at Lashly Glacier, 200, 201
at Mt. Crean, 196
on Mt. Gudmundson, 132-133, 134
on Mt. Ritchie, 145, 146
on Mulock Glacier, 138
personal characteristics, 71, 169, 174, 200
at Portal Mountain, 181, 182, 190, 191
Starlifter (C-141B), 31, 69-70
Starvation, 45, 46, 61, 101
Static electricity, 35, 208
Stonington Island, 10
Stratified lakes, 58-59
Stratigraphic measurements, 131-132
Suess, E., 3
Sulawesi (Indonesia), 160
Sunlight, 33, 38, 219-220
Survival gear, 68-69, 70, 102
clothing, 2, 24, 30, 37, 43, 68, 93, 98, 216
Survival training
Antarctica, 37-38, 39-41, 69, 72-73
deep-field, 69
Tekapo, 21, 22-28, 38, 69
Suter, K., 219
Swartz Nunataks, 150
Swedish Antarctic expeditions, 99

Tate Peak, 151
Taylor, T. Griffith, 4, 13, 58
Taylor Group, 50
"Teardrop," 223
Teeth, problems with, 23, 71
Tekapo (NZ), survival training, 21, 22-28, 38
Telephone service, 82-83
Temperatures
acclimatization to, 71, 213-214
before blizzards, 158
effect on tools, 118-119
extremes, 43, 75, 116, 118-119, 120, 122, 134, 135, 142, 150, 151, 194, 202
and food requirements, 99, 128, 213-214

inside tents, 122
pack ice, 76
Scott Base, 33
water, 36, 59
wind-chill factor, 82, 87, 96, 156, 194
Tennyson, Alfred Lord, 80
Tentacle Ridge, 111
Tents. *See* Polar tents
Terra australis incognita, 4
Terra Nova expedition (1910-1913)
accommodations, 27, 76, 78-80
deaths, 6, 7, 15
depot journey, 177
documentation, 6, 43
Far Eastern Party, 100
field cookery and food, 27, 43, 76, 177
fossil collections, 14-16
huts, 78-80
members of, xii, 11, 14, 27, 76, 193-194
Northern Party, 27, 76
planning and equipment, 6, 39
return party, 193-194
South Pole trek, 11, 39, 78, 129, 193-194
Terranes, 47, 199
Tetrapods, 161, 176-177
Thailand, field expeditions in, 159
Thelodonts, 169-170, 196
The Thing, 125
Thomson, Keith, 160-161
"Three Degree Depot," 129
Toilets, 70, 162-163
Topography, ancient environments, 132
Tourism in Antarctica, 20-21, 32, 221-222
Trace fossils, 10, 50, 60, 105-106, 107, 109, 134, 152-153, 170
Tractors, 28, 39
Training. *See* Survival training
Transantarctic Expedition, 23
Transantarctic Mountains, xii, 9, 125, 189
age, 13
expeditions, 7, 16, 29, 44, 48-49, 50, 65, 69, 108
fossils, 16-17, 33, 46-47, 65
geology, 96
ice- and snow-free areas, 57-58
view, 29, 70, 88-89, 128, 149
Trewin, Nigel, 108, 152
Trigonotarbids, 103
Trilobites, 46, 47
Trinity College, 215
Tumblagooda Sandstone, 152-153
Turnbull, "Moses," 48
Turner, Sue, 198

"Ulysses," 80
United States
Air National Guard, 66, 84
Antarctic Expedition, 42-43
Antarctic Program (USAP), 8, 68, 81
South Pole base, 69
University of
Hobart, 68
Kansas, 46
New England (Australia), 107
Oslo, 12
Sydney, 6
Texas at Austin, 47
Western Australia, 102, 214

Vanda Station, 38, 50, 57, 58, 60, 62, 63
Vandals, 63
Vega Island, 86
Ventifacts, 61
Vertebraria, 16
Vialov, O., 107
Victoria (Australia), 118, 120, 122, 146-147, 161, 170
Victoria Glacier, 59
Victoria Land, 8, 32, 47, 89, 118, 123, 149, 180, 182, 205
Victoria Valley, 50, 60
Victorian University of Wellington, 8, 23, 108
Volcanoes, 6, 32, 33, 34, 44, 56, 132
Von Bellinghausen, Fabian, 5
VUWAE
on Alligator Ridge, 168

expedition 15, 8, 143, 146, 147, 176, 179, 193
fossil collections, 136, 142, 189, 193, 196
parties, 8
VXE-6 Squadron, 44, 177-178, 180
birth of, 30-31
female crew, 69, 70
missions, 69
recon flights, 200
retrieval of abandoned gear, 205-206
safety record, 31, 84
South Pole landing, 31
US Air National Guard replacement of, 66, 84, 137
XD-03 recovery, 65-66

Warren, Guyon, 179, 198
Warren Ranges, 112-113, 125-126, 135, 137, 142, 147, 148
Waste disposal, 2, 35-36, 51, 58, 70, 81, 136, 222
Water temperatures, 36, 59
Weddell Sea, 13
Weddell seal, 51, 52, 73
Wegener, Alfred, 4
Western Australian Museum, 67, 87, 108, 156, 163-164, 215-216
Whales and whaling, 5, 16, 220, 221
White, Errol, 179, 198
White Desert, The (book), 23, 80
White Island, 34
Wigwam RNZAF base, 31
Wild, Frank, 82, 129
Wilkins, Hubert, 80
Williams, Vaughan, 1-2
Williams Air Field ("Willy Field"), 35, 66, 71, 205
Wilson, Edward, 6, 7, 11, 15, 43, 79, 110, 129, 188
Wind-chill factor, 82, 87, 96, 156, 194
Wind
accidents related to, 147
katabatic, 95, 192, 219
power generation, 78, 219-220
psychological effects of, 168
shelters, 145
speeds, 57-58, 79, 82, 92, 95, 97, 145, 147, 160, 164, 166, 194, 195
Windy Gully Sandstone, 62
Wintering over in Antarctica, 5-6
Byrd's solo, 215
food, 43, 54-55
huskies, 51-52
McMurdo Base, 37
Scott Base, 33, 72
Snow Hill Island, 13
women, 10
Women in Antarctica, 9-10; *see also specific individuals*
Woofer wood, 41
Woolfe, Ken, 49, 92, 104, 108
Woolly mammoths, 11
Worst Journey in the World, The (book), 43, 188
Wright Valley, 50, 57, 58, 60-61, 62, 64, 65, 84, 86

Xenacanthus, 123-124, 189, 194, 197

Young, Gavin C., 8, 9, 120, 124, 145, 147, 177, 160, 176, 179, 185, 189, 193, 194, 197, 198

Zanclorhynchus, 36
Zircon formation, 2